ENVIRONMENTAL PUBLIC RELATIONS MANAGEMENT

Principles, Strategies, Issues & Cases

Ikechukwu E. Nwosu, Ph.D
Vincent O. Uffoh, M.Sc

First Published 2005 by
Institute For Development Studies
University of Nigeria, Enugu Campus
Enugu

ISBN: 978-2409-74-X

DEDICATION

This book is dedicated to all lovers of the Environment; to the UNEP, & FEPA; and to our wives and children, born and yet to be born.

CONTENTS

PREFACE AND ACKNOWLEDGEMENT

Right from the Adamic age and up to this day, man has taken his environment for granted. Man has neglected or not paid adequate attention to his environment. His major interest, even as far back as the stone age, is to exploit or "use" the environment, mainly for satisfying his lower, higher and other needs. This exploitation was exacerbated by the emergence of modern civilization or modern development. As a result, the environment is now in serious danger. So also is the humankind or man. Is man aware of these dangers? If he is, what has he done to prevent the serious threats created for humankind by the overexploited and over abused environment? The truth is that many more individuals and groups all over the world seem not to know or are unaware of the great dangers posed by the *environmental time bomb*, than those who know or are aware.

This situation is made worse by the fact that even those who know or are aware of the environmental time bomb do not seem to care enough about it to take necessary actions to stop it from exploding. To all these insensitive or ignorant individuals and groups add the very many individuals and groups that, for one reason or the other (including religion and culture), tend to be fatalistic or resign themselves to fate, with the strange belief that God or nature or the ecosystem will take care of itself as it has always done since ages.

All these underscore the need for more proactive interventions by environmentally concerned individuals and groups in the world today and for them to redouble their efforts aimed at increasing people's knowledge and awareness of environmental issues and problems which will help to bring about widespread change of attitudes, opinions and behaviours about the environment; as well as ultimately translate to sustained and systematically positive environmental management, actions, programmes and projects that will help to save humankind from the impending environmental catastrophe. This book is a step in that direction.

This book, *Environmental Public Relations: Principles, Strategies, Issues and Cases Management,* is about man and his physical environment. It is about balanced or sustainable development. The book is on environmental public relations, public relations and social marketing and the mass media. It is focused on how various aspects or ramifications and strategies of modern public relations management can be gainfully and systematically employed in helping to manage environmental issues and problems in the contemporary world.

In addition and also very importantly, this book is written to help fill a gap in the public relations literature. It is meant to serve as a basic definitive book that will help the growth of knowledge and practice in environmental public relations (EPR) management. The book does not pretend or even want to be the last word on the new but very important area of EPR. It instead, hopes to serve as a catalyst for more of such authoritative and research-based books on environmental public relations.

The book also recognizes and is driven by the guiding truism that environmental management is essentially a multidisciplinary field of research and practice that calls for various inputs or insights from various professions and professionals, or occupations and occupational areas.

Environmental Public Relations Management: Principles, Strategies, Issues and Cases is in five-parts, each part is focused on a broad area of EPR that is handled from various perspectives by different chapters that make up the part. It is, on the whole, made up of 12 chapters that handled various aspects of EPR and EPR management or applications as well as six informative appendices.

PART ONE of the book, which was written as the book was about to go finally to bed (i.e. final printing) is focused on the analysis of various aspects of the "environmental monster" known as the TSUNAMI. It serves as the prologue to this book.

PART TWO of this book consists of four General Introductory Overview chapters that set the scene or lay the foundation for what will be coming in the rest of the book. Chapter one offers an explanatory, definitional, functional and historical overview of the principles and practice of orthodox and modern public relations. It also introduces the reader to the basic and leading models in general public relations management or practice as well as the meaning and application of the concept of *"Publics,"* public opinions and public attitudes. The raison d'etre for this opening chapter is that you have to understand the theory and practice of public relations before you can understand its off-shoot and specialized area known as environmental public relations (EPR).

In chapter Two, therefore, we handled the task of defining and explaining this new and specialized area of public relations known as EPR. In doing this we explained and analysed the EPR concept and practice from two perspectives, and used an earlier model propounded by one of the authors of this book, Prof. Ikechukwu Nwosu (2002), which is known as the Holistic Four Triangles Model of Environmental Public Relations (EPR), to further expand our readers knowledge of EPR principles and applications. Under the concepts of holism, ecology sustainability and EPR, we discussed the various management and marketing strategies, which the EPR manager must know and apply correctly before he can achieve maximal results. Chapter Three focused on another 1996 PR and EPR model designed by Prof. Ikechukwu Nwosu, the RICEE Model, and offered an indepth analysis of the model as well as how best it can be employed in EPR management.

The last chapter in this overview part of the books, Chapter Four, was used to give the reader a deep insight into the environments of environmental public relations and to emphasize that the EPR manager should understand and periodically analyse the various internal and environmental factors, forces and actors that will be driving or determining his decisions and actions as a manager, or specialist. This chapter did not only offer the EPR manager, the methods for analyzing these environmental factors, but went further to identify and describe these key factors, actors and forces.

PART THREE of this book is made up of two chapters whose purpose is to expose the reader and the EPR manager to an indepth analysis of the key or leading environmental facts, problems and issues in the contemporary world which were examined from various perspectives and with deliberate focus on the situation in Nigeria as a representative case example of the developing countries' situation. For sure, every chapter of the book touched upon, in some way, these environmental facts, issues and problems. But we decided to use the two chapters in this section of the book to give some deeper insights on these problems, facts and issues. We titled Chapter Five, "Some Vital Facts that the EPR Manager Should Know About Environmental Management, while we title Chapter Six "Environmental Degradation and Environmental Public Relations: The Past, Present and Future.

PART FOUR which deals with Environmental Public Relations Management provides the reader and EPR manager many practical application guides that can be used in EPR implementation. These are contained in three chapters that dealt deeply with many specific and concrete or practical EPR strategies, techniques, models and guidelines. Chapter Seven in this part, for example, had the leading and indicative title of Essential EPR Management Implementation Models and Strategies. Chapter Eight, on the other hand, handled the all-important need to disseminate adequate and reliable information to people on environmental issues, problems and projects and offered four strategic environmental information dissemination models that were later configured into an Integrated Model of Environmental Information Dissemination. In Chapter Nine, we introduced the EPR manager to the critical area of environmental impact analysis (EIA), with intentional focus on its methodological approaches.

PART FIVE of this book which concludes it, is a very important section of the book. It offers in three chapters, three original demonstrative and illustrative researches and case studies on environmental public relations (EPR) management. Chapter Ten, for example, is an elaborate sample survey case study that systematically analyses and demonstrates that the RICEE Model (Nwosu, 1996) can be applied systematically in environmental public relations management. Chapter Eleven offers an insightful report of a content analysis-based study of media images of environmental issues and problems in Nigeria. and finally, in Chapter Twelve, we offer a "hands-on" learning or group discussion case study on environmental incidents and public relations strategy, with nine guiding questions for tackling the key issues and problems of environmental management posed by the case study. The book ends with six appendices on the policy and regulatory aspects of environmental protection in Nigeria, some in full, some as excerpts.

We acknowledge with thanks the contributions of our colleagues, and friends and relatives who contributed in one way or the other to the completion and publication of this book, including the many authors, researchers, organizations and experts, we made reference to their work and publications. We give special thanks to the Federal Environmental protection Agency of Nigeria (FEPA) for supplying us with their various documents, excerpts of which form the various appendices to this book. To our readers, we say, a happy and enlightening reading, while we give all the praise and adoration to the Almighty Creator for giving us the creative and physical energy, the resources and good health to produce this book.
To the Lord be the glory.

Professor Ikechukwu E. Nwosu, Ph.D
Dean, Faculty of Business Admin.,
University of Nigeria (Nsukka)
Enugu Campus, Enugu, Nigeria

Vincent O. Uffoh, M.Sc.
Public Relations and Environmental Management Consultant
C/O Faculty of Environmental Studies, Enugu State University of Science and Technology, Enugu, Nigeria

PART ONE

- IN THE BEGINNING …
- BEFORE THE FEAST
- RAISING THE CURTAIN

****

PROLOGUE TO
THE
BOOK

****

PROLOGUE
The Tsunami Disaster As A Global Environmental Warning-Bell And An International Public Relations Masterpiece

As this book on Environmental Public Relations Management was about to be finally published by its new publisher, the former publisher worked on the manuscript for almost two years and failed woefully, the Asian Tsunami, the world's most destructive environmental catastrophe so far, which affected mainly 12 countries in the coastal areas of southern Asia, struck. It even affected parts of Africa like Kenya, Somalia, Tanzania and Madagascar.

This Tsunami which killed hundreds of thousands of human beings, animals, plants and other natural elements, also wiped out many towns, houses, roads and other structures. In terms of dollars and naira loss, the multi-billions of total money-loss may never be fully determined.

Strangely and ironically also, the Tsunami disaster occurred on December 26, 2004 when people from various parts of the world as well as the indigenes of the areas most seriously affected, were in a festive mood as they were enjoying the Christmas and Boxing Day holidays. The primary lesson we should all learn from the Tsunami disaster therefore, is that it can occur anywhere and any time as a national, regional or global environmental cataclysm of no small magnitude. Every part of the world, including Africa and Nigeria (the main geographical focus of this book), should have a plan of action to deal with this catastrophic emergency, wherever and whenever it occurs. The present world response to the victims of the Asia Tsunami which is commendable but rather slow an somewhat belated in many cases, should be also seen as an exemplary masterpiece in international public relations, as well as a powerful support for our plea in this book for greater application of public relations strategies in environmental issues and problems management.

It is significant to note that the Asian Tsunami was able to do all the havoc it wreaked because it was an admixture of a powerful *undersea earthquake* of 9.0 Richter scale or magnitude which triggered ferocious *tidal sea waves* that surged at a speed of 800 kilometres per hour and took their greatest toll on lives and any other thing on its way, in the Asian countries of Indonesia, India, Sri-Lanka, Thailand, Malaysia, The Maldives, Myamar and the Seycheles Island. Aceh, a remote town in the Indonesia region of Sumatra was the worst hit because it was near the *epicenter* of the 9.0 earthquake and so was instantly hit by the generated tsunami which claimed instantaneously a total death toll of more than 85,000 people in this single area alone, and destroyed at least 60% of the town Aceh.

And as if God was particularly angry with people of the Sumatra region of Indonesia, more than 1,000 people were again killed and a lot of structures destroyed in the Nias Island in this area when another major earthquake struck on March 28, 2005 (on Easter Monday), just three months after the devastating Christmas period tsunami of 26th December, 2004 (CNN, March 28, 2005). Fortunately, this earthquake was not accompanied by the deadly tidal waves of the December tsunami. Apart from the Aceh town, the December 2004 tsunami also caused heavy loss of lives and property in Thailand's Khao Lake beach, a tourist town in which according to many media reports, corpses floated like rafters, bloated and mangled bodies danced to the macabre rhythm of the waves; and most of the 770 recovered bodies belonged to different nationalities (including tourists and the indigenes).

One positive fall-out of the Asian Tsunami debacle is that it has most stupendously and even rudely re-awakened world leaders, experts and even ordinary citizens of mother earth to the inevitable need to keep looking for solutions to the many global environmental problems facing us today, including the prevention of the preventable ones. This could be part of the reasons why the long – outstanding KYOTO PROTOCOL or CONVENTION on Global Warning and the pollution of the atmosphere, finally came to full force in February, 2005. It is sad and unfortunate though that till date one of the world's greatest polluter countries, the United States of America (USA), has refused to sign that United Nations Convention on the environment, very sad indeed. It was even more disappointing and painful to watch President George Bush defending the U.S.A.'s refusal to sign the Kiyoto Convention in a recent Cable News Network (CNN) special programme that warned that if the world does not take concrete steps to check the current global warming and ozone-layer depletion trend and regain already lost grounds on this impending global environmental doom, the consequences will become permanently irreversible and sweepingly destructive.

The Tsunami disaster also lent more significance to the award of the 2004 Nobel Prize on the Environment to Dr. Mathai Wangari. The award was announced just a few months after the Tsunami debacle. Sixty-four years old Wangari thus became the first African woman to win the Noble Prize. She is from Kenya and has been appointed into the Kenyan Government as a parliamentarian after the award. This Noble Prize was given to her for her vigorous and successful advocacy and works on environmental preservation in Kenya, which made her to clash many times with the Kenyan Government. The implications of Dr. Wangari's Noble Prize award and successes for African scholars, policy makers, women, environmentalists, media, public relations and other environmental advocates

4

or activists should be sharply obvious to anyone who makes out time to read this book on public relations and the environment.

It is important to point out at this juncture that tsunami is not as new as some people may think. It is indeed on record that since 1819, about 40 tsunamis had struck the Hawiian Islands. The Table below (Table One) also shows in a graphic manner, the tsunamis that have occurred in the world in the past 100 years, when and where they occurred as well as their strengths or magnitudes as indicated by the areas they covered in kilometers and feet. The December, 2004 Asia Tsunami is the biggest in the last ten decades, in terms of the areas it covered and destructions it brought about as well as its speed of 800 kilometres per hour.

Table One

Tsunamis Caused by Earthquakes and Landslides in the Last 100 Years

S/No	Dates	(m)	(ft)	Location
1.	August 27, 1883	9	30	Java Sea
2.	October 6, 1883	10	34	Alaska
3.	June 15, 1896	38	125	Sanriku, Japan
4.	September 10 1899	60	197	Gulf of Alaska
5.	September 30, 1899	12	39	Banda Sea
6.	June 26, 1917	11	36	Samoa Island
7.	March 2, 1933	29	96	Sanriku, Japan
8.	April 1, 1946	35	115	Aleutian Islands
9.	May 22, 1960	25	82	Chile
10.	March 28, 1964	70	230	Gulf of Alaska
11.	October 6, 1979	3	9.8	Nice, France
12.	September 1, 1992	11	36	Nicaragua
13.	July 1, 1993	5	16	Japan
14.	June 3, 1994	60	197	Eastern Java, Indonesia
15.	July 17, 1998	15	49	Papua New Guinea

Source: *Microsoft Encarta Encyclopedia, 2002*

More Facts on the Tsunami Phenomenon

It seems necessary for us to use this point in the prologue to this book, to provide further facts and figures about the tsunami phenomenon to our readers, especially those who heard about it for the first time in December 2004 and might not have had the time or opportunity to get more facts on it, as well as those who may never have heard of this intriguing and deadly phenomenon at all. Firstly, tsunami is originally a Japanese word "that means harbor wave". It is used today however, in scientific and environmental management circles to refer to seismic sea waves generated by an undersea earthquake or possibly an undersea landslide or volcanic eruption. The tidal waves created by a tsunami undersea earthquake or landslide is comparable in structure to the concentric waves created by an object dropped into the water, but is understandably billions of times much more powerful and destructive.

According to the *Microsoft Encarta encyclopedia* (2002 edition), a tsunami can have wavelengths or widths of 100 to 200 kilometres (i.e. 60 to 120 miles) long and may travel hundreds of kilometers across the deep ocean reaching speeds of about 725 to 800 kilometres (about 450 – 500 miles) per hour. Upon entering shallow coastal waters, the wave which has been only about half a meter (a foot or two) high out at sea, suddenly grows rapidly. When the wave reaches the shore, it may be 16m (50ft) high or more. Tsunamis, which usually owe their tremendous strength to the great volume of water affected are capable of obliterating coastal settlements, as happened in the December 2004 Asian Tsunami. Tsunamis should not be confused with cyclones, hurricanes, storms or storm surges, which can be dangerous or destructive, but are not as destructive as the tsunamis.

The Tsunami and The Animal-Versus-Man Sixth Sense Imbroglio

It will be serious oversight to end this prologue to this book without drawing the readers' attention to the still-raging controversy that I like to describe as the Animal-Versus-Man sixth sense Imbroglio, which was generated essentially or at least escalated by the December 2004, Asian Tsunami disaster.

This controversy arose from and is fuelled by the simple question. Do animals have a sharper sixth sense than human beings? According to media reports on the Asian Tsunami, wild animals living in the areas affected by the Tsunami (listed earlier), migrated away from these areas before the Tsunami could deliver its deadly blows and destructions that affected mostly human beings who are considered to be very intelligent or at least more intelligent than animals, and who should have sensed danger earlier than the animals and so run away even much earlier. Does this incident and the differential reactions of men and animals to the Tsunami suggest in any way that animals may have shaper sixth sense or instincts than human beings. Maybe, may be not.

In fact, while some experts and ordinary people may draw conclusion from the above – sketched or reported scenario and so answer "Yes" to the controversial question raised above, some researchers in the USA, without necessarily or directly referring to the Asian tsunami, would probably answer "No" to the question, based on their research findings. According to them, they have identified an area of the brain in human beings that provides us with a sixth sense and may enable us to unconsciously take steps to steer clear of danger. They further state that this part of the brain provides a sixth sense that serves as an early warning system by monitoring our environment and weighing up the possible consequences of our actions or inaction.

According to a Daily Mail of London science reporter, Robin Yapp (Daily Sun, Feb. 24, 2005), these research findings could help explain why it was that Sri-Lankan tribes people somehow sensed the impending danger of the Boxing Day Tsunami and fled to higher ground in time to save themselves. So, from his report we are informed that at least some human beings (tribesmen) also used their sixth sense and moved away from the Tsunami catastrophe, just like the animals. But the fact that it was tribesmen (not city or urban men) who are usually considered by many (especially white racists) to be closer to animals than city or urban men, that used their sixth sense to move away from the Tsunami danger, raises a whole set of questions again, questions which can only be answered by hard data provided by solid research.

This book, *Environmental Public Relations Management,* did not attempt to answer these questions raised by the animal-versus-Man sixth sense imbroglio because they are far beyond its scope. But it definitely raised and answered many more mind-boggling questions that have more day-to-day survival applications and implications for the continued survival of animals, men and other living things in their environments today and tomorrow. The Tsunami warning bell has tolled and is still ringing. We as professionals/experts and leaders do not have to wait for another tsunami before taking the crucial decisions and actions that will save mother earth. That is essentially what this book is all about, even with its focus on the role of public relations in the management of environmental issues and problems.

Prof. Ikechukwu E. Nwosu Ph.D.,

Source: The Nigerian Journal of Communication Vol. 4, No. 1, 2005

PART TWO

****

GENERAL
INTRODUCTION
OVERVIEW

****

Chapter One

Public Relations: An Explanatory, Functional And Historical Overview

Introduction and Definitions

This book is essentially focused on the understanding and result-oriented application of public relations as an essential strategy for managing environmental issues and problems in contemporary global scene and in a development context. It also deals with the possible influences and implications of environmental issues management for contemporary and future public relations practice. It seems right, therefore, to offer early in the book an explanatory and functional overview of public relations. This is because we shall not be able to use or apply a tool or strategy unless we first understand it and its proper uses.

Explaining Public Relations

What then is a public relations and what is it not? Public Relations is as old as human civilization and human environment, in many forms and shapes. It has been used in many ways, for many purposes and in various contexts or millieux. For our purpose in this book, however, we are interested mainly in its contemporary form, contexts, meanings and applications. And the current meanings and applications can be traced to the "grand – daddy" of modern public relations' works; the works of Edward Bernay, which indeed laid the foundation for other works in modern public relations by other authors, scholars, managers and experts. Bernay who is on record as teaching the first structured public relations course at the university level (the University of New York, U.S.A.)wrote the firs serious book in public relations in 1923. The title of the book, *"Crystallizing Public Relations,"* is still a tight and very useful pointer to what public relations is and what it does.

This is because, it will become crystal clear as we go through this chapter or when we get to its end, public relations is essentially about positively and systematically using actions and communications to influence people's attitudes, opinions, beliefs, interests and behaviours in a given or desired direction (e.g. adopting of good environmental behaviours) as well as building lasting CREDIBILITY and REPUTATION for individuals and corporate entities that include profit or non-profit organizations and even nations, states, local governments or communities.

In fact, Bernay's later book, *Engineering of Consent* (1986) underscores this point because it sees public relations as basically involving the attempt to use information, persuasion or appropriate actions to "engineer public consent for a cause, idea, activity or programme".

And in the context of this book, public relations is definitely a veritable tool that can effectively be used as information, persuasion (i.e. communication) and actions to engineer or bring about generally accepted public consent, attitudes, opinions interests and behaviours for positive environmental management as a cause, an idea, an activity, a programme or a project.

Other definitions of modern public relations have used different words and emphasized different things about it; but they are not too far from the above definitions. And the professionally accepted ones were more than 600 when we took the last count in 1996 (Nwosu: 1996). In this and some other chapters of the book, we shall select and offer the leading ones among these definitions, especially those that are most relevant to the management of environmental issues and problems.

Thomas C.O., and his associates (1998: 548) defined public relations as "a marketing and management functions that focuses on communications that foster goodwill between a firm and its many constituent groups." We particularly like this definition because of its emphasis on the marketing management and communication functions or dimensions of modern public relations, which makes it an indispensable tool for managing environmental issues, problems and projects in modern society, developing or developed.

We shall touch some more on these three vital aspects of public relations in this chapter and other parts of the book. Even though the action dimension of public relations is not directly stated in this definition, it is understood and must not be missed or neglected by the reader in any result-oriented public relations efforts (i.e. environmental public relations management). This is because modern public relations emphasizes action and places it above information and communication. This is why some of us have defined public relations as involving planned and systematic actions that are properly communicated to the right publics (Nwosu, 2001).

The 1978 Mexican statement states that:

> Public Relations is the art and social science of analyzing trends, predicting their consequences, counseling organization, leaders and implementing planned programmes of actions which wills serve both organization and the public interests.

Furthermore, Cutlip, Center and Broom see Public Relations as "the management function that identifies, establishes and maintains mutually beneficial relationship between an organization and the various publics on whom its success or failure depends (1984).

From these definitions, it is evident that public relations involves deliberate, planned and sustained programmes of action and regular two-way communication between an organization, (government, business etc), individuals and their publics that are aimed at positioning the organization as *credible and reputable*. It can also be defined as the art and science of building and sustaining a credible reputation for any organization (i.e. "Reputation Management", Nwosu, 2000). According to (Sam Black, 1990), the fundamental purpose of public relations is to establish mutual understanding based on truth, knowledge and full information.

We have to restate at this juncture that the one major bane of public relations practice in the developing countries like Nigeria has remained the wrong understanding, conception, perception and therefore the wrong application of public relations in many quarters or contexts, sometimes by even managers or top government officials who one would expect to know better and so perform better. It is exactly this situation that made Ikechukwu Nwosu (1992 and 1996) to posit that one method of knowing what public relations truly is, as a profession or modern management practice, is to also know what it is not. This thesis or submission led him to identify and divide the various definitions of public relations into:

(a) The Nonsensical definitions of public relations;
(b) The commonsensical definitions of public relations;
(c) The Technical or professional definitions of public relations (Nwosu, 1996).

According to Nwosu (1996) the nonsensical definitions of public relations are those definitions that are based on abject or utter ignorance, shallow knowledge, misperception or pure mischief on the part of the definers or describers. To him, such nonsensical definitions and description of public relations include the ignorance-based and derogatory ones that wrongly present or project public relations as cover-up, window dressing, bribery, gimmicks, propaganda, mere image making (as opposed to image or reputation management), employing empty-headed beautiful young ladies or 'macho' young men and putting them in the front office to smile and be nice to visitors of the organization, mere publicity or just another advertising or marketing functions.

They also see public relations managers or describe them as mere lobbyists, busybodies, praise singers, their masters' voices or megaphones, errand boys, and personal assistants who do such menial or demeaning jobs as procuring prostitutes for their bosses and board members as well as carrying the bosses' briefcases to and from the airport, and help their bosses' wives to shop and take their children to school (Nwosu, 1996).

Public relations as a corporate management function is of course, none of this. If it were to be, then it would never have been recognized and recommended the world over as a management strategy that can be used to manage various environmental issues and problems, as well as a tool for performing the many other functions of public relations, which we shall summarize later in this chapter.

Nwosu (1996) went further to describe the *Commonsensical definitions* of public relations as those that are based on common sense, and went ahead to give some examples of such definitions. To him, those definitions come very close to what modern public relations really is all about, yet they should not be seen as substitutes for the professional or technical definitions of public relations. *Commonsensical definitions* include the ones that described public relations as doing good and getting credit for the good done; doing good and ensuring that you got caught in the act; doing unto others as you would have them do unto you; and many others. Nwosu (1992) even formulated his now-popular *commonsensical definition* of public relations as "making friend, keeping friends and working with friends to achieve your objectives".

According to Nwosu (1992) the technical or professional definitions of public relations "are those definitions that are based on proper and deep understanding of the theory and practice of modern public relations as a management function and those that are based on thorough research, experience and deep thought on the role of public relations in modern corporate entities in business, government and non-business/non-government sectors". These are the definitions the environmental public relations practitioner should study, understand, adopt and adapt in applying public relations in solving different problems in the human environment or the ecosystem. Such definitions will no doubt include the ones we have offered in this chapter and others that will be offered in other chapters.

Functions of Public Relations

The key and minor functions of public relations can be deduced from what we have written so far in this chapter about it and the various definitions we offered pointedly. However, we can state that the functions of public relations include the following: Corporate and other communications, persuasion, attitude and opinion management, information management, counseling; corporate strategy and policy formulation, corporate image and reputation management, event management or event marketing, protocol functions, community relations, corporate social responsibility, employee relations, marketing support, financial public relations, research and evaluation, international public relations (IPR), planning, publications, corporate advertising, publicity, publications; audio-visual productions (e.g. documentary films), public enlightenment, corporate identity, issues management, crisis management and many other functions that are related to the above list (Nwosu, 1992 and 1996).

The major functions of public relation includes, of course ENVIRONMENTAL PUBLIC RELATIONS which this entire book is all about which involves systematic and planned application of public relations theories or principles and practices in managing various projects or programmes.

We must stress at this point that for these public relations functions to be properly and usefully employed, individual groups or organization that wants to apply them must recognize public relations as a top management

16

function in which the public relations executive or director reports directly to the chief executive officer (CEO) who is regarded by public relations experts as the "Number One Public Relations Officer", by virtue of his position. He does not necessarily have to be a public relations expert but he must be public relations literate, public relations friendly, public relations sensitive or conscious, public relations driven and public relations compliant.

Secondly, for these functions to be usefully applied or gainfully employed, all the workers or staff in the organizations wanting to use public relations must be public relations literate and conscious, starting from the smallest messengers, to the top-most managers. Without this cooperation from everyone in the organization, the public relations director or executive and his team cannot optimally achieve set public relations objectives for the organization or project. This is because they are not magicians or spirits who can be everywhere all the time. In other words, public relations cannot give us the desired results unless we practice team building, team sustenance, and team participation (Nwosu, 1996).

Some Leading Public Relations Models

As Grunig and Todd (1984) have rightly pointed out, public relations as a social science or management science is still engaged in a vigorous effort to formulate generally acceptable and universally applicable theories. What exists now in some reasonable abundance are public relations models. Some of the existing public relations models will be summarily identified and discussed in this section. In doing this, we shall concentrate on those models we believe can be gainfully applied in environmental public relations management.

The RACE model is perhaps, the earliest and most widely cited model in modern public relations practice. It is an operational guidance model that also helps to explain the public relations process. In it, R stands for Research, A stands for Action, C stands for Communication and E stands for Evaluation. It stresses the importance of adopting a systematic approach in carrying out public relations projects as well as the great relevance of research and evaluation in modern public relations practice. Most importantly, it underscores the fact that public relations practice must consist of a well thought out, planned and executed actions that are well communicated.

The Transfer Process Model by the late Frank Jefkins, (1982) is another leading widely known and widely applied model of public relations. Its major message or lesson is that public relations strategies can and should be used in any attempt to change over or convert people's attitudes from their negative states to positive states. According to the model, public relations strategies can be used to change Hostility to Sympathy; Prejudice to Acceptance; Apathy to Interest and Ignorance to Knowledge. The Transfer Process Model can be applied at the entire corporate/organizational image and reputation management level and at the specific project/programme management level. This is what makes it specially invaluable in the application of public relations principles and strategies in managing environmental issues and problems (environmental public relations) management or practice.

The RICEE model propounded by Ikechukwu Nwosu (1996) is another public relations model we will identify and briefly discuss in this chapter, since details on it will be offered in different chapters of this book. It is worthy to note that the RICEE model was originally designed and recommended by Nwosu (1996) as a public enlightenment model in managing environmental issues and problem. It has of course gone beyond this scope and has been applied in other areas of public relations management and practice. A case study was used in this book to demonstrate the RICEE's model application in Environmental sanitation management (see Case Study Number One in Chapter Ten). In this model which will be better explained later, R stands for Research, I stands for in-formation, C stands for Communication, the first E stands for Education; and the second E stands for Evaluation.

There is also the Operational Matrix or IPCM model propounded by John Marston (1985), which we consider very useful in environmental public relations management (EPR). It introduces the idea of applying marketing strategies along with public relations strategies in trying to achieve public relations and other related objectives involving persuasion and changing of attitudes and opinions. For environmental public relations and social projects, this model marketing will be seen to be most appropriate and most useful. In this model I stands for information gathering analysis and dissemination, P stands for Public affairs (i.e. identifying critical public affairs issues, problems and facts and factoring them into our public relations projects), C stands for Communication and M. stands for Marketing.

Finally, we will introduce here, briefly discuss and strongly recommend the four closely related models propounded by James Gruning and Todd Hunt (1984). These are the Press Agentry/Publicity Model, the Information Model the Two-way Assymetric model and the two-way Symmetric model. The strength of these four inter-related models lies in the fact that they can collectively serve as an effective analytical tool for examining the development of public relations as a management function, field and discipline in which the press agentry/publicity model is regarded as the earliest state and the two-way Assymetric model is regarded as the latest most advanced level of development in PR practice. They can also collectively serve as operational or practice-oriented model that will help the practitioner or manager using these models to know which one of them to use as a strategy at any particular time and in every particular context. For example, some environmental management problems in public relations may jut require the application of press agentry or publicity aspect of the four models, another EPR project might just require the less-recommended two-way Assymetric or the more-recommended two-way Symmetric model which is regarded as the most advanced and most effective model of public relations practice in modern times (Nwosu, 1996).

The Concept of Publics

As you go through this chapter and most of the book you will find the word or concept of publics all over. What is public? Publics are a public relations professional jargon or concept that is used to describe the targets (individuals or groups of people) of public relations actions, policies, communications, decisions and projects or programmes. It is always in the plural form. It is comparable to the concept of *market* in marketing, which refers to human beings or people with money to spend and readiness to spend it on your products or services, if you work hard on them enough. The concept of publics is comparable to what we describe as *receivers, audience or destination* of our messages in general communication theory and practice or in advertising. It is also comparable to what experts in mass media studies will call the masses or folks in traditional or folk communications media (Nwosu, 1990).

Technically, the concept of publics can be defined as everyone interested in an organization or is affected by that organization's policies, decisions, interests, communication or projects and whose needs, wants, actions, opinions, attitudes can affect the organization, and is consequently interested in the organization and differ from one organization to the other. One way of further understanding the publics of an organization is to see them as the Internal and External stakeholders of the organization. The stakeholders vary from organizations to organizations, of course, (Nwosu, 1996).

There are many ways of identifying the publics of an organization. The most popular is to group them into two: INTERNAL AND EXTERNAL PUBLICS. This is easy to understand. The next method is to divide them into PRIMARY, SECONDARY AND TERTIARY PUBLICS in public relations projects, incidents and events management. The primary publics are those that need the first attention (e.g. in a fire incident). The secondary publics are those that require and deserve secondary attention, action and communications. The tertiary publics are those that require last attention, (not least) communications and actions. In managing any environmental disaster and other crisis (e.g. in the oil industry), this second method of classifying publics will be most appropriate (Nwosu, 1996).

Another method used in classifying the publics of an organization is subdividing them into BASIC PUBLICS and SPECIAL PUBLICS. Again, this is simple to understand. The only thing to note is that the basic or fundamental publics for target groups can be internal or external. So also can special publics who require and indeed deserve special public relations actions, communication, attention policies and decision from the organization. Furthermore, we should note that basic and special publics are not always necessarily fixed or permanent in PR management scenarios.

Public Relations, Public Opinions and Public Attitudes

From our discussions in the previous sections of this chapter with regard to the meaning and functions of public relations, it can be deduced that public relations seeks to elicit goodwill, support for and build mutual understanding between individuals, groups or organizations and their various publics. It can also be deduced that public relations achieves its objectives through actions and two-way integrated and systematic communications, based on truth and full information, as opposed to lies, half-truth deceptions, mis-information, disinformation, buck-passing and other propaganda techniques. Public relations also achieves its objectives by applying its tools, strategies and techniques that are appropriate for the circumstances under consideration or the task/problem at hand, including environmental issues and problems. All these are deliberate, systematic and sustained not adhocishly executed. It should also be clear, at this point, that essentially, every public relations strategy is targeted at the relevant publics with a view to encouraging them to develop positive opinion and attitude toward an issue or organization or changing their opinions, and attitude favourably towards an issue or organization, if their previous opinions and attitudes were both unfavourable. Thus, the publics' attitudes, opinions, views, acceptance or rejection of any laudable programme, individual, institution or organization is a crucial factor towards determining the success or failure of such programme or organization.

Against the above background, it can be said that public opinion and attitude are critical theoretical concepts, which the public relations executive must understand and must be able to influence, to the ultimate advantage, towards achieving his public relations campaign and other objectives.

In essence, there is no chance of a public relations programme succeeding without a deep understanding of the bases and nuances of public opinion. Little wonder Howard Stephenson and Philip Leseley (1978) defined public relations respectively as the art of "convincing people that they should adopt a certain attitude..." and "all activities and attitude intended to judge, influence and control the opinion of any group or groups of persons in the interest of any individual group or institution".

Public opinion has been defined and described by many authors but most of them revolve around the fact that public opinion represents a consensus, which merges, over time, from all the expressed views that cluster round an issue in debate and that this consensus exercise power. Hennessy (1981) defines public opinion as the complex preferences expressed by a significant number of persons on an issue of great importance. On the other hand Cutlip et al (1984) split up the two words that make up public opinion for better understanding. They define publics as collective noun for a group of individuals tied together by some common bond of interest and sharing a sense of commonness. Furthermore, they view opinion simply as the expression of an attitude on a controversial issue. Furthermore, Canfield and Moore (1973) define public opinion as an expression of a belief held in common by members of a group or public on a controversial issue of general importance.

Public opinion gets its power through individuals, who must be persuaded and organized. However, Cutlip et al (1984) maintain that people act on the basis of "the picture in their heads" rather than in accordance with the reality of the world outside and to understand and influence them one must take into cognizance these "pictures in their heads. They went ahead to enumerate those things constituting the pictures as; personal factors, environmental factor, culture, the family, religion, school, economic class, social class, race, age, personal motivations; and the role of the group. Also, Edeani (1993)observed that scientific research has shown that, in terms of public opinion, there are at least three main categories of publics: The mass public; the attentive publics and opinion making publics. The mass public are at the lowest cadre and they are opinion holders who have little opportunity or inclination to express in any meaningful way the opinions which they hold. They therefore rarely bother to participate in the opinion making process. However, the opinion making public who constitute about 2% of the population are the people who form, hold, and express opinions on most issues of public interest in view of the positions they hold, their wealth, their educational attainment or their social standing. Hence once this opinion making class has been won over in a particular PR campaign, the public sooner or later joins the camp.

On another note, Festinger (1957) in his theory of cognitive Dissonance gives another dimension of public opinion. He states that individuals tend to avoid information dissonant or opposed to their own point of view while they tend to seek out information that is consonant or in support of their point of view. He maintained that the people whose attitudes and opinions can be influenced most readily are those who have not yet formed on opinion – the so-called passive or neutral members of the public. Hence, an understanding of this Festinger's theory would go a long way towards helping the PR campaign to win support and goodwill for his organization through persuasive communication, or build reputation and credibility for it.

Yet on another note, Rokeach (1996) defines attitude as a relatively enduring organization of beliefs about an object or situation. He suggests that attitudes change is not merely a function of attitude toward an object but also of attitude towards a situation. Wiebe (1953) thinks that opinions affect attitudes according to the demand of casual situations. But having adapted them, opinion appears to become ingredients in the constant and gradual reformation of attitudes. Hence the attitudes of individual citizens are the raw material out of which a consensus develops. Therefore, influencing individuals' attitudes is the prime task of the PR practitioner.

The construction of attitude is widely conceived as having three components, viz; the cognitive, which is the knowledge component; the affective or feeling component and the psychomotor or behavioural component (Edeani, 1993). Edeani believes that in order to change an existing attitude, the three components must undergo changes in some ways. However, he emphasizes that the cognitive component is relatively easy to change, while the affective is most difficult because it is attached to emotions.

Public opinion is commonly regarded as an expression of an underlying public attitude, and this relationship casts public opinion in the status of a vehicle for the communication of public attitudes. Thus, the PR person must understand that the relationship between public relations on the one hand and public opinion and attitude on the other is simply reciprocal in character. Therefore, the most effective public relations campaigns are those that recognize these facts about public opinion and attitudes and so endeavour to tailor such campaigns to appeal to the publics or audience in question.

Brief Historical Perspective

While modern public relations as a management function can be said to be relatively new in the developed countries and even newer in the developing countries, the point must be made that public relations is indeed as old as human civilizations or history in both the developed and developing countries. Its history, can indeed and has been traced as far back as the creation of man (Adam) by God and his decision to create later a woman

(Eve) to *relate* to him as a companion in that *beautiful original human environment* described in Genesis Chapter One as the Garden of Eden which we have lost long ago. Environmental and EPR specialists and others are working hard to ensure that bits and pieces of what it left of the long-lost and destroyed Garden of Eden (mother earth) are preserved for the human beings of today and tomorrow. Even God himself was very interested in being understood by Adam and Eve (human beings) and still is. And public relations in its most basic form is about building and sustaining understanding among individuals and groups without which there will be no relationships, or at the best uneasy, strained and temporary relationships that will crack and die a natural death. This indeed is in a nutshell, what public relations is all about.

However, many other factors contributed to the development of Public Relations as a field of study and a management practice. These factors include the Renaissance Movement of 1440-1500AD and the Reformation and Counter Reformation between the Catholic and the Protestant churches led by Martin Luther. For example, the awareness created by the Renaissance offered people the right to debate issues, question the kings' orders and rulings, ask questions, investigate into nature as well as discuss together matters of public interest. Then came civilization, which emanated from the above mentioned struggles and served as a solid factors for the birth of public relations practice.

Other factors that contributed to the development of public relations include the Firs and Second world Wars; the American War of independence which gave her the lead in almost all things, including public relations. And then French Revolution which is a liberating uprising against feudalism and a clarion call for freedom and peace, understanding, harmony, fairness, truth and all the ideas which modern public relations can rightly be associated with.

However, Scot Cutlip, Allen Center and Broom (1984) traced the profession back to the ancient Greeks who conceived the idea of "Public

Will", as well as to the Romans who introduced and popularized the expression, "Vox populi,Vox Dei", meaning that the voice of the people is the voice of God. In the words of Cutlip, Center and Broom:

> Effort to communicate information to influence viewpoints or actions likewise can be traced from the earliest civilizations. Archeologists found a farm bulletin in Iraq, which told the farmers of 1800BC how to sow their crops … much of what is known of ancient Egypt, Assyria, and Persia was recorded effort to publicize and glorify the rulers of the day. Much of the literature and art of antiquity was designed to build support for kings, priests and other leaders ... Rudimentary elements of public relations can be found in the history of ancient India. The functions of the king's spies included keeping the king in touch with public opinion. They also championed the king in public and spread rumors favourable to the government.

They went further to state that in England, public relations was introduced many centuries ago when the kings maintained Lord Chancellors as "keepers of the king's *conscience*".

These "Chancellors" served as intermediaries between the kings and the people. They were always consulted by the king and could be said in modern times to be public relations practitioners, because their opinions were sought by the kings before they made decisions or policies or took action on matters of *public interest;* just like present day public relations executives.

In Russia, Catherine the Great was "a genius on publicity" who was reported to be an expert on publicity because she saw publicity as an essential instrument in the art of government (and the formation for modern public relations). She was reported to have written her speeches and manifestoes in a way that ensures they made the fullest appeal to her people.

She is also on record as using public relations as a veritable agent of public opinion formation and an instrument for mobilizing public support.

In the U.S.A., the history of modern public relations can be traced to 1441 or thereabout when press agentry and printed publicity developed and flourished. Some experts, however, traced the origin of public relations to the mass letter-writing campaign to gain public acceptance and sway the press in faovur of the United States constitution organized by such revolutionary figures as Alexander Hamilton, John Madison and John Jay (Bitner, 1977). Another important milestone in the development of modern public relations in the U.S.A. is the invention of telephone in the 1800s by Alexander Graham Bell.

But perhaps the most important landmark in the growth of modern public relations in the U.S.A. is the fact that the United States government used public relations as a key strategy in mobilizing her citizen's support for the First World War. The United States Committee on Public Information, headed by George Creel, provided the platform for this effort, and their public relations efforts further strengthened the success of World War II public mobilization efforts. The Second World War public relations efforts were planed and executed by the U.S. War Information Office which was headed by Elmer Davies. Many organization and agencies in the U.S. later started their own public relations and publicity practices and in that country today almost all information departments operate a public relations unit. In the private sector of the United States of America, Ivy Lee, Edward Bernays, Lein Baxter and Clem Whitaker laid the foundation for the business of public relations consultancy that is flourishing in the U.S.A. today.

In Nigeria, the origin of public relations can be traced to pre-historic times from the traditions and traditional practices of Nigerian leaders and citizens, even though they did not really understand or describe those practices as what we know today as public relations. For example, the leaders who were known as either Kings, Obas, Emirs, Ezes, Obongs, Amanyanabos and other such traditional nomenclatures, had many ways of communicating with their citizens and mobilizing support for developmental activities, wars, agriculture, environmental and village square or streams cleaning, and many such activities. And as we pointed earlier, public relations involves mainly the systematic use of communications and actions to achieve individual, group or corporate objectives like building mutual understanding between an institution, or a person and his publics (in this context, the king/leader and the citizens).

They even had very sophisticated ways of communicating and relating to individuals and groups outside their immediate communities in peace and war times (e.g. inviting friends to their festivals or soliciting the support of allies for inter-ethnic wars). Also, many of their celebrations and festivals were public relations events in many respects. In fact, they even had methods of publicity or advertising their goods and services for sale or barter. There are many others of such actions and communications in pre-historic Nigerian societies that can aptly be described as the roots or foundations for the development of public relations in Nigeria.

However, modern public relations' development in Nigeria can be traced to the Second World War era and immediately after that war. The British colonial government noticed a palpable and growing dissatisfaction among the Nigerian citizenry and had to introduce measures to check this. Since they could not get positive results from their coercive efforts, they decided to introduce and also use persuasion and information management, which are basically public relations strategies. This led to the British colonial government's appointment of information officers who combined the duties of government publicity and public relations and setting up in 1894 a public relations department in Lagos with units in Northern, Eastern and Western Nigeria.

Some multi-national companies like the United African Company (U.A.C.) also appointed public relations officers who helped to further consolidate the development of modern public relations in Nigeria. The few professionals among them later got together and formed the Public Relations

Association of Nigeria (PRAN) which is indeed the root of the Nigerian Institute of Public Relations (NIPR) that is now empowered to regulate or control public relations in all its ramifications as a chartered professional body. But it actually took until 1990 before the NIPR was given legal teeth by Decree Number 16, which is now an act of parliament or the National Assembly. That was during the presidency of the NIPR by Mr. Mike Okereke who worked for U.A.C. for 23 years as public relations manager and actually helped to truly professionalize public relations practice in Nigeria. he did not only popularize the then already existing NIPR code of ethics,But helped to formulate and implement some new NIPR bye-laws to further regulate the practice in Nigeria, embarked on large-scale membership drive, started the public relations journal and public relations newsletter, but successfully ran many training workshops in public relations all over the country as well as started for the first time the NIPR Professional Certificate and Diploma Examination series in Public Relations under the Educational Advisory Board, which is now known as the Education and Standards Board of the NIPR.

It is also significant to note that it was Mr. Mike Okereke as an NIPR president that worked with the University of Nigeria Nsukka (at Enugu Campus) to start in 1992 the first-ever full-fledged M.Sc. and Postgraduate Diploma Programme in Public Relations in Nigeria, and Africa. That programme is very much alive, kicking and plans to start a Ph.D. Degree programme in public Relations in the 2003/2004 academic year. Other past presidents of the NIPR include the founding president, the late Dr. Sam Epelle, Chief R. Akinyele, Chief Bob Ogbuagu, Chief Kanu Offornry, Alhaji Sabo Mohammed and Chief Jibade Oyekan. The current president is Mr. Sofri Bobo Brown who is now working hard to further modernize and expand the frontiers of public relations practice and profession in Nigeria.

Conclusion

Let us conclude by simply saying that public relations as a profession and management practice has gone a long way in the world and Nigeria. it is now a full-fledged profession the world over, a recognized management practice and full-blown academic discipline with a fast developing body of knowledge and research, which we have merely tried to introduce and explain in this chapter in an encapsulated form. There is for sure, much more that can be said about the public relations profession such as the recent emphasis or focus of the profession on REPUTATION MANAGEMENT (Nwosu, 2001). But there is also a limit to what can be said about a whole profession and academic discipline in a single chapter. The reader will definitely get some more knowledge on various aspects of public relations in different chapters of this book. But he is advised to procure and read other good books in public relations some of which are listed in the references contained in this book.

Very importantly and for the purpose of this book, the reader should also procure and read anything he can on this new and important area of environmental public relations (EPR) management (2004). It does not matter whether these EPR materials are textbooks like this one, book chapters as in the next chapter of this book, professional or technical journal articles, magazine and newspaper article – read them and become wiser by drinking from the well of knowledge and inspiration which they really are.

Chapter Two

Environmental Public Relations (EPR): A Definitional and Explanatory Overview

Introduction

Over the years, man has exploited nature and his environment to his advantages paying little attention to sustainability of the environment upon which his very existence depends. It was on this note that some environmentalist maintain that world leaders, albeit without intending to, have created a civilization that is headed for destruction. They went further to warn that, we either learn to control our growth in economic activity and population towards environmental sustainability or nature will use death to control it for us. *Time International Magazine* (1991) in its publication entitled *"Learning How to restore the Wilds of Eden"* captures the picture most vividly by asserting that *"if the fate of human depends on nature, the fate of nature irrevocably and irretrievably rests in human hands"*. Hence the key to man's survival on earth is sustainable development.

According to the World Commission on Environment and Development (WCED), in the 1987 Brundtland report, "Our Common Future", sustainable development is defined as the development that meets the needs of the present without compromising the ability of future generations to meet their own needs...

> It is a process of change in which the exploitation of resources, the direction of investments, the orientation of technological development and institutional change are all in harmony and enhance both current and future potentials to meet human needs and aspirations.

But to secure a worthy environmental legacy both for ourselves and for future generations, we must find ways to reconcile humanity more satisfactorily with the natural systems and the environment upon which all human life and civilization depend. This is based on the understanding that the natural systems of which we are a part have an intrinsic worth, transcending narrow utilitarian values.

However, the ideal sustainable environment is far from being realized in most developed and industrialized countries, not to talk of the developing nations within which our discourse and research in this book is located.

This is why we shall propose in this chapter the adoption and implementation of the various principles, practices of environmental public relations (EPR) as a veritable strategy for sensitizing and persuading individuals and groups in our various societies to change their negative attitudes and behaviours towards environmental issues and problems and adopt more environmental friendly behaviours and attitudes. In adopting this EPR option, we also strongly recommend the adoption and use of the RICEE and other relevant models of public relations we discussed in previous chapter. We should recall that the PR RICEE Model was propounded by Ikechukwu Nwosu (1996) essentially as a model formulated for planning and executing public enlightenment campaigns aimed at changing the opinions, attitudes and behaviours of people toward realizing sustainable environments. We shall give future details on this model in the next chapter. According to Nwosu (1996), "the adoption of the EPR Public enlightenment and Public Education opinion is informed by the fact that the days are gone when any major issue of public policy can be formed without widespread public support". Moreover, "if a proposal cannot be presented in a form in which it can gain support of the general public, prospects for its success are dim indeed (Nwosu 1996). The campaign against environmental degradation is not an exception in this regard. And therein lies the great importance of EPR.

The full realization of the obvious consequences of leaving the environment to degenerate led to the popularity of the new concept of environmental communication and environmental public relations (Nwosu 1996). To underline the importance, the United Nations came up with the United Nations Environmental Programme (UNEP) a body charged with global environmental monitoring and regulation. Equally to fulfill its own national obligation, Nigeria set up The Federal Environmental Protection Agency (FEPA), to monitor and coordinate the efforts of individuals and corporate bodies towards a harmonious environment. Morestill, various state governments have instituted their own state environmental agencies. For example, in Enugu State, there is the Enugu State Environmental Protection Agency (ENSEPA). In the same line, various non-governmental organizations are contributing their quota in the bid to maintain a healthy environment. But the questions still remain: what strategies are those bodies using? Have those strategies shown a remarkable result, which goes to justify the huge chunk of resources being invested on them?

It is still evident that many industrial outfits do not care for the negative effect of their activities (like industrial pollution) on the environment? Equally, improper sewage disposal, indiscriminate liquid and solid waste disposal, sullied vicinity, contaminated potable drinking water and other environmental pollution still look us in the face, showing that more still needs to be done in order to avert or at least seriously minimize the scourge and plagues of polluted environment such as epidemics.

It becomes obvious that Environmental Public Relations could be applied to communicate more effectively, elicit goodwill, influence people's opinion, judgement and consequently behaviours and attitudes towards the environment.

Environmental Public Relations (EPR)

It seems necessary to proceed by more pointedly defining and explaining the concept of environmental public relations for easier reference, perusal and understanding.

Simply put, environmental public relations is a specialized area of public relations and reputation management practice that focuses on how best to apply relevant public relations principles, practices, strategies, techniques, models and tactics in any effort to properly manage environmental issues, problems and projects and so ensure the achievement of sustainable development objectives at the communal or local, national, regional and global or international level. But it is much more than this as we shall see later. It is a specialized public relations concept or strategy, which grew out of our observed utter neglect of the deliberate, planned and systematic application of relevant public relations and reputation management strategies at various levels in past attempts to manage environmental issues and problems.

EPR is a specialized public relations and reputation management strategy that also arose from our observation that even those environmental management managers or experts who make any effort at all to apply public relations and reputation management do so lackadaisically, haphazardly, adhocishly, half-heartedly, tokenisticaly and without making any real effort to first acquire some good working knowledge of public relations and reputation management; or employ some trained public relations executives or managers to work with them; or use the services of professional public relations consultants, in order to properly apply public relations strategies in the management of environmental problems, issues and projects.

34

EPR also grew out of our observed gap in knowledge in the area of public relations literature or studies and practice, and our desire to contribute towards the filling of that gap with the hope that more public relations scholars and experts will start doing more research and publishing more technical journal articles and books that will help to create abundant and reliable body of knowledge in this specialized area of public relations (EPR) that touches on all aspects of our existence on this earth, no matter where we are because, for example, the air we all breath, the water we all drink and the sunshine we all enjoy, know no boundaries.

If public relations and reputation management is a relatively new management function and is consequently facing many misperceptions, misunderstanding, misapplications and even poor recognition and underutilization, we should expect that the much newer environmental public relations (EPR) is bound to face all these negative factors and a lot more. The problem though is that EPR as a specialized area of public relations study and practice, unlike many other areas of regular public relations and reputation management practice, has to do with life and death because of its focus on critical and dangerous problems in the ecosystem or the human environment that can massively destroy man, animals, plants and indeed the entire earth as we know it today. Even though crisis and issues management; as a specialized area in modern public relations and reputation management, has to do with life and death (Nwosu, 1996), it is not as crucial as EPR, at least for two reasons.

Firstly, most issues and crisis management problems are usually geo-specific and their consequences hardly ever pose life and death danger to the entire earth or a very high percentage of the world's population. Secondly, strictly speaking, a major focus of EPR is issues and crisis management because many environmental problems today have gone beyond mere localized problems to global issues, and if not properly managed at the issues stage, as research on crisis management have shown, they will sooner than later graduate into full-blown crises (Nwosu, 1996), in line with the Crisis Life Cycle discussed in detail in the book, *Public Relations Management* by Ikechukwu Nwosu (1996, 104-122). In fact, many of the environmental issues in the world today are already at the crisis level or stage and many United Nations agencies, nations, governments and non-governmental organizations (NGOs) are already spending billions of Dollars in tackling them as global challenges. All these go to clearly and strongly demonstrate the importance of EPR.

The Other Side of EPR: The Flip Side of The Coin

Under this subtitle we will further expand our understanding of EPR in this chapter by giving other definitions and explanation of it, in addition to the ones we have given already. We shall also design and describe an EPR model that will help us to more clearly understand what it is or what it does. All these will be aimed at expanding the one-sided and unidirectional definition and explanations of the concept of EPR, which we intentionally gave earlier because that is easy to understand, and also because that is really the major focus of or approach to EPR we want to emphasize in this chapter.

In offering these other sides or dimensions of the meaning of EPR, our emphasis will extend to the inescapable fact that environmental issues and problems themselves influence EPR decisions, communications and activities or practices. These constitute what we described above as the flip side of the coin in EPR management and practices, which we should bear in mind and be guided by. They also underscore the need for us to adopt a multi-dimensional, wide-spectrum and holistic approach to the understanding and application of EPR principles and strategies.

Towards an Expanded and More Appropriate Definition of EPR

We can define EPR as a set of public relations management activities that are aimed at creating public acceptance and competitive advantage for any corporate entity by convincing its stakeholders (domestic and international) that its policies, products and activities are totally

36

harmless to the environment or at least environmentally-friendly, especially when compared with the policies, products and activities of their competitors. This definition sounds right and appropriate. But it is narrow, simplistic and incomplete. It is at best only 50% of what EPR is or should be, as we shall see later. It is also similar to the intentionally simply definition of EPR we gave earlier.

It can also be said to be opportunistic and too heavily focused on or directed towards the achievement of tactical advantage, rather than the pursuit of the strategic change needed to ensure the achievement of broad and specific environmental and societal development goals and objectives. Furthermore, it is dangerously too close to the now discredited lip service and lopsided or one-side approach by companies (.especially commercial or profit oriented ones) to dealing with environmental problems and issues. As one critic of this discredited approach rightly observed, "selling and *Public Relations* activities with a green theme were rife, but this had little connection to customers' (stakeholders') needs or to the realities of the environmental impact of the products (and policies) involved" (Peattie, 1995).

A publishing company may, for example, write on all its publications the following promotional and environmental-friendly stunt or statement: "The Publisher's policy is to use paper manufactured from sustainable forests" (Pitman Publishing Company, London). That is well and good. But we need to go beyond this slogan or statement, get inside the company or organization, critically examine its policies, products and activities and most importantly, procure research insights from its internal and external stakeholders, before we can give it the credit of being a truly environmental-friendly company. That is the only way.

It is imperative therefore, that for us to get closer to the true, expanded and appropriate meaning, definition and explanation of EPR we must relate it to or situate it in the very complex and inseparable socio-cultural, political, economic, technological, physical (ecosystem) and other environments in which organizations (profit and non-profit) and their stakeholders or public exist and interact with each other. Based on all these, we shall operationally define EPR in this chapter as follows: EPR is a holistic management process and specialized area of public relations management that is responsible for identifying and anticipating (forecasting) the ecosystem needs, interests, policies, public activities, issues and programmes of any corporate entity (e.g. company, country or state parastatal). And also implementing sustainable programmes of public relations actions and communications that will *reconcile them* with the ecosystem needs, interests, expectations, demands, activities, problems and policies of the stakeholders of that corporate entity (domestic and global). This is to ensure sustainable development through environmental-friendly corporate policies and actions, as well as the perception and acceptance of that corporate entity, its products, services, policies and activities as environmental-friendly by all the domestic and global stakeholders of the corporate entity and the environment.

This is an intentionally long definition that is supposed to, at the same time, explain the various dimensions of EPR, especially its essential two sides that will always remind the EPR manager that while he is using public relations and other related strategies to ensure that his organization is environmental-friendly. He should never forget to factor in these environmental problems, actors and issues into his organizations' policies, projects, programmes, plans and activities because, whether he likes it or not, these environmental problems, issues and actors do have significant or serious influences on his organization's EPR policies, activities, performance and overall rating on the environmental-friendliness or management scale in the mind and eyes of environmental actors and the general public (domestic and global). This might be described as the dualism or the dual nature of effective EPR management.

Towards A Holistic Model of EPR

To further expand our understanding of EPR principles and applications, we shall offer and explain below our EPR model (Nwosu, 2003: 12 and 2004: 45-59).

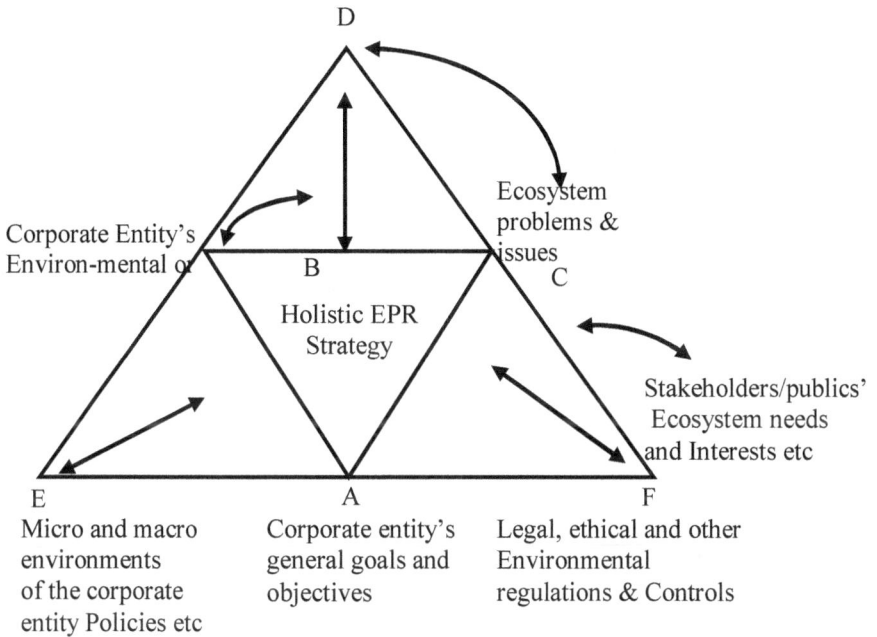

```
                              D



                                        Ecosystem
                                        problems &
Corporate Entity's                      issues
Environ-mental o|       B                    C

                    Holistic EPR
                      Strategy
                                                   Stakeholders/publics'
                                                   Ecosystem needs
                                                   and Interests etc

  E                     A                   F
Micro and macro     Corporate entity's   Legal, ethical and other
environments        general goals and    Environmental
of the corporate    objectives           regulations & Controls
entity Policies etc
```

Fig. 2.1: The Holistic Four Triangles Model of Environmental Public Regulations (EPR)

Source: Nwosu, Ikechukwu E. (2002: 12).

As shown in Fig. 2.1 above, the inverted triangle ABC encapsulates or encloses the recommended Holistic EPR strategy that forms the hub of the Holistic Four Triangle Model of Environmental Public Relations. In this inverted triangle, the A tip or segment represents the corporate entity or organization concerned or whose environmental performance is being examined or handled with the EPR strategy. The A tip also shows that we should start by identifying, understanding and critically examining how that organization's general policies, goals, objectives, corporate strategies and plans are influenced by or are sensitive to environmental issues and problems as well as how they can influence the organization's ecosystem or environmental policies activities etc. which are shown in tip B or segment B of the ABC inverted triangle. Triangle ABC shows and emphasizes the need to reconcile always the ecosystem or environmental needs, policies, activities and interests of that organization with the environmental needs, expectations, interests, demands and activities of its stakeholders or publics, all in a holistic manner.

Also, as shown in Fig. 2.1, the normal (uninverted) bigger Triangle DEF envelops the EPR triangles (ABC), and the D part or tip of that triangle houses the Ecosystems or general environmental problems and issues which the EPR manager and his company must manage well and which also impinges on the organization's overall policies, strategies and activities. The E tip or segment in Triangle DEP houses the micro and macro environments of the organization (e.g. political, economic, social, technological environments) which influences the EPR policies and activities of the organization, and which the EPR manager must effectively manage in order to achieve his organization's EPR objectives. And lastly in the DEF triangle is the F segment or section which draws our attention to the legal, ethical and other influenced by the EPR management decisions, actions, communications, policies and programmes of the organization concerned. The three arrows inside triangle DEF which all point to triangle ABC (the EPR triangle) and which have two pointed ends, go to underscore the holism or interactive philosophy or relationships among the various parts of the model which should make them systematically and perpetually interdependent.

Holism, Ecology, Sustainability and EPR

Another look at our operational or expanded definition of EPR above will show that we worked with the concepts of holism, ecology and sustainability in trying to come up with that definition. We therefore need to tightly explain and discuss them here to further bring out their meanings and utility in understanding and practicing EPR management.

The origin of the concept of holism can be traced or credited to J. Smuts who in his 1920+ book titled *Holism and Evolution* stated that nature's evolution and progressions towards ever more complex forms and organizations is driven by a tendency to form wholes that are more than sum of the parts, through the process of ordered groupings, and that it is only through the appreciation of the whole entities and interactions among them that life can be understood. Smuts' 1920 holism concept must have directly or indirectly influenced Ludwig von Bertalanffy's (1968) General Systems Theory because there are a lot of similarities between their thoughts and conclusions. It is also on record that it was as from the 1960s and 1970s that management experts like Peter Drucker (1973) and i. Unterman (1974) started promoting the idea or concept of holism in management studies and practice. All these gave birth to the holistic management strategy.

But as Ken Paettie (1995:30) rightly pointed out, the holistic management strategy "was originally prescribed for companies to tackle the problems of internal control and coordination. In order words, it had an internal closed system, production orientation". And as the operational definition and the model of EPR we offered in this chapter clearly shows, holism in EPR management must involve taking an internal and external open-systems marketing orientation to achieve the desired objectives. This involves seeing the organization concerned not just as a holistic corporate entity in techno-economic terms, "but as part of a socio-environmental *ecological* system" (Paettie 1995:31).

Marketing Strategies in EPR

Since EPR is holistic management process, its practice must extend far beyond the normally recognized boundaries of public relations theory and practice, especially in terms of its philosophy, strategies, techniques and mode of application. For example, EPR must be more proactive and interactive than the run-of-the mill public relations practice or management. It must also put into use many more marketing and management strategies than general public relations practice. For example, it must be guided by the principles of SOCIETAL MARKTING which is essentially a marketing philosophy that emphasizes the corporate social responsibility (CSR) or community relations (CR) principles of ensuring that we use EPR to make sure that our companies or other organizations contribute appreciably or reasonably to any society or community in which they operate or do business in, so that they will not be seen as distantly, uncaring or uninterested landlords whose only interest is to collect their rents (profits) at the end of every month and smile to the banks while the citizens of these societies and communities suffer neglect and poverty.

Related to the concept of societal marketing is the concept of SOCIAL MARKETING, which is very often confused with societal marketing. It is more of a marketing strategy, which is used in the marketing of not-for-profit product, services, ideas and institutions, such as environmental issues, and problems. Also important is the concept of DEMARKETING which refers to the marketing strategy that can be used to stop a consumer from consuming harmful products like cigarette and marijuana, or engaging in unwholesome behaviours like having multiple sex partners, as well as polluting the environment with all kinds of waste materials.

In addition, the EPR manager is expected to embrace the principles and strategies of RELATIONSHIP MARKETING which essentially involves building and sustaining close and lasting relationships with customers, distributors, suppliers and other important or special stakeholders of the organization to maintain loyalty, cooperation and lasting respect and understanding. This can be quite an expensive strategy; but the pay-off is worth it, especially when properly applied and managed. The EPR manager should not have problem with this because relationship marketing grew out from the good old marketing-support function of public relations known as CUSTOMER RELATIONS or customer care/services (Nwosu, 1996:5-8).

Management Strategies in EPR

With regard to management, the EPR manager will also benefit immensely from leading and relevant management strategies for maximal success in his job, while applying the holistic approach or the concept of holism. The understanding and application of STRATEGIC MANAGEMENT principles will come very handy to him. It will help him to properly align his EPR activities to the organization's vision, mission, policies, philosophy and objectives, as well as make correct strategic decisions and choices. But even more importantly, the application of strategic management principles will help him to go beyond the internal environments of his organization and respond well to the changes posed by the external environmental factors (including the ecosystem itself) by evolving, implementing and supervising the EPR strategies that will ensure resounding performance, internally and externally. Of course, he can also easily and proficiently carry out other critical strategic management activities like strategic planning, environmental scanning ,monitoring ,evaluation ,strategy audit and SWOT or TOWS analysis

Adopting and applying the holistic management principles will also help the EPR management to always adopt a HUMAN RESOURCE MANAGEMENT (HRM) strategy that is human-centric, motivational and unexploitative. These will help him to properly empower his EPR team and encourage other workers in the organization to respond appropriately to the various publics and stakeholders of the organization in order to mobilize, team up and work with them to achieve the organization's environmental management objectives. In like manner, the EPR manager and his team will help to ensure that an environmentally-friendly TOTAL QUALITY MANAGEMENT (TQM) strategy is not only installed but practiced for the satisfaction and full involvement of the "internal customers" and "external customers" of the organizations towards tackling the organization's qualitative and maximal eco-performance in the organization. This brand of TQM can be referred to as TOTAL QUALITY ENVIRONMENTAL MANAGEMENT (TQEM).

Furthermore, the EPR manager must put to practice and preach the gospel of MANAGEMENT RE-ENGINEERING which involves a total and radical restructuring or redesign of both human, material, financial, processes and the corporate culture or the way things are done in the organization concerned. Re-engineering requires a lot of vigour, discipline and accountability for it to be beneficial and successful, because if wrongly applied, it can be very disastrous generally and painful to the organization's stakeholders. So, the EPR manager has to use it with care in order to reap from its benefits and avoid its pains. This is especially so, because generally speaking environmental management with its redesign or radical attitudinal behavioural policy and overall change-orientation "shares the 'clean sheet' mentality of re-engineering and approach which has the potential to bring companies remarkable benefits and extreme pain (Paettie, 1995:32).

Ecology and EPR

What of the Ecology concept? What does it mean and how does it relate to EPR management. We have been using the ecological system or ecosystem concept throughout this chapter alongside with the concept of the physical environment,, which is virtually synonymous with it. In trying to explain the ecology concept we shall adopt the very apt and lucid four-points description of the concept offered by B. Commoner (1972) in his book entitled *The Closing Circle*. The first point to note in understanding what ecology is, is that "everything is connected to everything" in nature. This idea makes the ecologist to see the environment as a "web of life" in which a change to one strand, unit or element (e.g. an organization or community) can have repercussions through the whole network or chain. The global ecosystem or environmental web will therefore, collapse, if one or all of its strands or units is continually abused or destroyed.

The second point posited for the understanding of the concept of ecology is the point that "everything goes somewhere". Physicists tell us that matter and energy cannot be destroyed; so they must go somewhere else. This is why any pollution, industrial waste, toxic material or other such anti-environmental matters constitute clear and present danger to one and all, at the individual, corporate, national or global levels. This is also why the EPR manager should embrace and use this particular principle or point in ecology, in planning and executing his EPR campaigns, and in advising his organization on environmental issues, problems and policies.

The third outstanding ecology explanation point or statement we must note is that "nature knows best and can always balance itself". We should therefore, desist from all activities that tend to change the "balance of nature", for economic profit, development or other purposes. This is no doubt, a strong or powerful lesson for the EPR manager in doing his job. This is the *balance of nature* idea or message of the science of ecology, which the EPR manager must learn. That message, lesson or point is that "there are no such things as free lunch". This means that the bill for every intervention may come decades later, and sometimes even to a different address, but environmental damages definitely have to be paid for by someone, somewhere, sometimes.

EPR and Sustainable Development

What about the SUSTAINABILTY concept and its relationships to and applications in EPR management? This concept has its root in the concept of sustainable development, which has its origin in the 1980 World Conservation Strategy, which was further, crystallized and more widely disseminated in the Bruntland Report (WECED, 1987). Sustainable Development has been defined as the development strategy that meets the needs of the present without compromising the ability of future generations to meet their own needs (WCED, 1987).

M. Jacobs (1991), identified the three key components of sustainability which we believe wills serve as appropriate guides to the understanding and application of the sustainability concept by the EPR manager and his team. The first component is *Futurity* which involves the adoption, in all we do, as EPR mangers, a long-term perspective which gives equal consideration to the needs of future generations and to our own needs, both at the individual and corporate levels, especially with respect to environmental protection or preservation.

The second component is *Welfarism,* which underscores, the need for the EPR manager, his team and us all, to always factor into our activities and policies a deep consideration of the benefits accruable to every individual in every society from these policies and activities. This will help to ensure improved quality of life for all, now and in the future.

The third component or element of sustainability is *Equity.* This calls for a concerted effort by individuals and organizations to endeavour to balance the distributions of economic costs and benefits between different communities, countries, regions, socio-economic classes, ethnic groups, religions, races and sexes in all that they do. This sense of equity and fair judgment will no doubt form a cardinal point in the EPR management policies and activities of any organization, for it to achieve any reasonable results or impact in managing environmental issues and problems.

Perhaps it is for these and many other reasons that can be given that a senior lecturer in Strategic Management at the Cardiff Business School, Ken Peattie (1995:33) has emphatically stated that:

> Sustainability is no longer an alternative policy; it is now a generally agreed principle of future economic growth and development. The international chamber of commerce's *Charter for Sustainability* has been signed by over 1000 of the world's leading companies, and sustainability is an

inherent part of the European Union (EU) policy, following the publication of the Fifty Environmental Action (Plan), Towards Sustainability.

We cannot agree more with the above statements and will indeed hasten to add that the whole of Africa through the AU (formerly OAU) have in 2001 and 2002 fully endorsed the principles and practical applications of sustainability as a development and environmental management philosophy.

This is contained in the historic policy document entitled New Partnership for Africa's Development (NEPAD), which will be discussed in some detail in Chapter Seven of this book.

Other EPR Management Strategies

We want to recommend at this point some other popular EPR STRATEGIES, which the EPR manager should use, depending on what the environmental situation calls for. The first is the OFFENSIVE STRATEGY, which requires the EPR manager to take the initiative with his PR actions and communications. This strategy is also called the PROACTIVE STRATEGY and is similar to what is known in marketing practice as offensive marketing strategy. Related to the offensive EPR strategy is what is described as the PRE-EMPTIVE STRATEGY. In fact, both of them are sometimes confused with each other. But the difference between them lies in the fact that we only use pre-emptive strategy anticipatorily; that is when there is an anticipated attack, danger, criticism (i.e. foreseen or expected). In offensive public relations strategy, there does not have to be a foreseen attack before you initiate an EPR campaign or project, programme or attack.

Then, there is the DEFENSIVE STRATEGY, which is used when a harm has been done to your organization, its policy or products. It is a retaliatory or corrective strategy for attacks you never foresaw before they occurred. And finally, we have the OPPORTUNITY STRATEGY, which is employed when the EPR manager sees a clear opportunity, after environmental scanning for example, and launches a campaign to take full advantage of it in the interest of its organization, clients, publics or products. The use of any of these strategies in EPR is made possible by the fact that public relations is a very flexible form of communication and in the area of corporate communications it is becoming increasingly important as a *strategic* tool and weapon (Paettie, 1995:226). In adopting any of the above EPR strategies, the usual potent *tactics* of modern public relations practice like press conferences, media appearances, interviews, speeches, press releases, corporate literature or publications, audio-visual material, ICTs, trado-rural media, information services, special groups sponsorships and donations, using special groups and others such tactics, will come very handy.

Conclusion

All that we have written in this chapter go to strongly demonstrate the great importance of EPR as a specialized area of public relations management, research and practice that should be given the urgent and serious attention it deserves, in our own collective interest as homo sapiens, nations, regions, and inhabitants of mother earth. For these and many other reasons that can be given, we strongly recommend that all academic departments, or programmes that teach or offer public relations (including marketing and mass communication and environmental studies departments) in universities and polytechnics should introduce at least one general and compulsory course in environmental public relations and reputation management (EPRM), plus one specialized EPRM course. Industries, government, UN agencies and non-governmental organizations should also closely liaise and work with these academic and research programmes or departments in the institutions of higher learning, to promote all kinds of interdisciplinary and multi-disciplinary researchers, training and education in all areas of the ecosystem, or the physical environment.

Professional chairs in EPRM should be established; specialized institutes and centres on EPRM and the environment in general should be set up in these institutions of higher learning and their existing programmes on the environment should be adequately supported financially, materially and in every other way. These are already happening in the developed countries of Europe and the United States of America. It should start happening in the developing countries of Africa, Asia, Latin America and the Middle East. That is the only way forward for everybody, organization, nations and the continents. It is the only hope sustainable data generation, strategies formulations, decision making and action taking that will ensure the survival of mother earth. After doing all these, we should leave the rest to the Almighty God who created mother earth and the other planets in the first place.

Finally, we like to start drawing the curtain in this chapter by charging the EPR managers in Nigeria, Africa and every part of the world to study and put to practice the tenets of the now-historic "IPRA-CODES OF PRACTICE on public relations and the Environment" produced by the International Public Relations Association (1992). Among other things, the code prescribed that:

1. Members shall not publicize or promote products, organizations, or services, claiming environmental benefits, unless those benefits are demonstrable in the light of current science and technology.
2. Members shall endeavour at all times to promote openness, which handles facts and concerns related to the environment and development.
3. Members shall seek to develop programmes, which counsel and communicate the benefits of balanced consideration of environmental, economic and social development.
4. Members shall provide a free flow of information within and through IPRA concerning environmental development issues on an international level.

National public relations institutes and other Professional Associations, such as the Nigerian Institute of Public Relations, should come up with and even more importantly, religiously enforce environmental protection codes of ethics and standards, with the IPRA Code above as guide. This will help their members to deliver resoundingly the results that various organizations and individuals are expecting from them in the crucial area of environmental public relations and public information management.

There are many of such publicly asserted or vocalized expectations from public relations practitioners and mangers, but we shall offer only two of such assertion to demonstrate our point or underscore this point. The first of such strong assertions came from no less a person than the director General/Chief Executive of the Nigeria's Federal Environmental Protection Agency (FEPA), Dr. Adegoke Adegoroye (1997:13). In his words,

> "It is in this respect, among others, that the greatest and true catalyst that is an inalienable and indispensable requirement for the realization of our stated vision is PUBLIC RELATIONS! Credible and prompt information to the public; resolution of inter-agency role conflicts; removing distortions in interpretation of environmental information available or disseminated to the public; presenting government views based on scientific and sociological facts; etc, are in my view, the areas of expertise of public relations officers".

What a powerful confidence/expectation! What a powerful charge! But to these assertions and charges we must add a very important missing link. That missing link is that EPR managers must also present adequately the community, society, public or peoples's views and facts on the environment to ensure a balanced information mix or flow pattern that will guide decisions and actions in all quarters on environmental problems and issues.

The General Manager (Public Affairs) of Chevron Nigeria Limited, Mr. Sola Omole, did not only underscore the expectation of many actors in environmental management that public relations managers and practitioners should lead the way in the crusade to protect the environment, but would agree with the missing link we supplied or added to Dr. Adegoroye's charge above. This is why he has written (Omole, 1997:11) that

> public relations, especially environmental public relations, must be built on credible, authoritative information. All of the brightly coloured and beautiful looking advertisements and printed brochures mean nothing, if they are not based on facts. Environmental public relations in our industry must be responsible for projecting and advertising (promoting) authenticated information as well as demonstrating and accepting accountability when there is a problem.

We agree completely with this very balanced and well-thought-out assertion.

Chapter Three

Indepth Analysis of The Ricee Model as a Practical Model of Environmental Public Relations Management

Introduction

One of the outstanding and original public relations models that have been developed and which will contribute towards the achievement of a safe and sustainable environment in the world is the RICEE model

Propounded by Professor Ikechukwu Nwosu (1996), as aforementioned in the previous chapter, this RICEE model is a public relations model, which we believe can be applied in the topical issue of environmental public relations management or how public relations can be applied in managing or controlling environmental issues and problems. The PR RICEE Model is a components of the model. As aforementioned, R, refers to Research, I refers to Information, C, refers to Communication, E, stands for Education and the last E refers to Evaluation (Nwosu, 1996).

For a comprehensive understanding of this model in all its ramifications, we shall try to explicate the five components that constitute it with a view to bringing into lime-light and better focus how each component contributes and compliments the others towards achieving integrated environmental enlightenment campaign objectives and consequent institution of sustainable healthy environment. When the model was first propounded, it was known as the RICE model, comprising – Research, Information, Communication and Education. But later, the need for evaluation was seen to be indispensable for the model to be really complete. Hence the fifth component was added. Thus, the model is now known as the Public Relations RICEE Model.

The Research Component

The first component of the model is Research. Research has been recognized as an indispensable tool in Public Relations, which is both art and science. Research is a process of arriving at reliable solutions to problems through the planned and systematic collective analysis and interpretation of data. It is a process of arriving at dependable knowledge. In Public Relations, we need data about our publics, their opinions, the intensity of such opinions, media use patterns and other characteristics. Raymond Simon (1980), conceived public relations research as planned, carefully organized, sophisticated fact finding, and listening to the opinion of others. He went ahead to identify six areas in which research plays vital roles for public relations work, viz:

* Provision of factual input for programming;
* Provision of input about public attitudes and opinion;
* Serving as an "early warning" system;
* Securing internal support for the public relations functions;
* Lubricating the public relations machinery.

Furthermore, Robert O. Carlson (1973) who was the one time Dean of the School of Business Administration at Adelphi University maintained that research in public relations is the educated counsel and experience that are brought to bear on the design of a new programme based on the cumulative experience of all Public Relations people. It is the feelings in the air of professional meetings and luncheons, and the prevalence of small talks with colleagues that alert us to some new concepts or development in our field. According to him, Research in Public Relations serves three functions. Firstly, it may simply confirm assumptions and hunches about the state of public opinion on an issue, a product or a company. It is a useful kind of back-up function, which may be analogues to the use of quality control system in the manufacturing business.

Secondly, research is used to clarify questions on which limited information is available or on which apparently contradictory data are to be found. For example, such studies can help determine if expressed attitudes are related to actual behavior. This implies that research can help sort out what people really mean when they say they like or dislike an organization (or an issue or programme) the reasons they cite for these feelings, and even the origin of the conceptualization on public relations problems. As it were, there is also an unintended bonus in conducting research.

There are various techniques in research studies, one of which is content analysis (Holsti, 1996; Nwosu, 1995; Babbie, 1975). Such research gives a pretty fair measure of the salience of the problem and often useful hints as to which aspects of it seem to be arousing greatest public interest. Equally, there is the survey research technique. This research comes in different forms as public opinion survey, profile survey, effectiveness survey, trend survey and panel survey. Equally there are the depth surveys and pre-tests which are tools for public relations and opinion research. For environmental issues however, survey research methods seem most popular. This is because survey research design deals with the practical application of the already standardized theories available in social and behavioural sciences (Ozongwu, 1992: 42; Osuala, 1992; Olakunori,1997; Kerlinger, 1973; Dominick, 2000; Ikeagwu, 1998).

In concluding survey research, we have to decide the survey objectives, determine the target groups and determine what degree of accuracy is required for our purposes. Equally we have to bear in mind that all sample surveys are subject to sampling error. Also we have to note that four ingredients are essential. They are: the sample, the questionnaire; the interviewers and the analysis (Gallup, 1997:3).

But this notwithstanding, content analysis can complimentarily be used with survey method depending on the particular environmental issue in question and the prevailing circumstances. It can also be used alone.

Information and Communication Components

On the other hand, with regards to environmental degradation, information which is the second component of the PR RICEE model, ensures a reliable system for disseminating regular, adequate and appropriate information to the relevant publics as regards current pollution (Nwosu, 1996). Information is a matter in which everybody has a stake. Everyone yearns to be informed about issues. The new information technologies are equally making it better to disseminate information faster, easier and in better forms. Yet, on another note, communication is another indispensable ingredient of the PR RICEE model.

According to Ikechukwu Nwosu, communication is a process of information, ideas and opinions exchange within, between or among individuals, groups, organization or nations (usually made up of human beings) in a social or societal context (Nwosu, 1996). Philip Lesly, (1973) would maintain that the ability to communicate is such a basic part of the human experience that it makes possible everything that distinguishes man from the rest of creation. The ability of one person or group to deal with other groups through communications processes is integral to the entire social nature of the human species.

Lesly, (1973), went further to assert that albeit public relations encompasses much more than communication, the essence of public relations is in the broad definition of the term "Communication". Communication incorporates sensing the state of rapport or lack of it with the publics involved; interpreting that state in terms of the group or organization and their objectives; assimilating the implications of this interpretation and adjusting the posture and thrust of the group or organization accordingly and transmitting those thoughts; and equally receiving feedback to help evaluate the effectiveness of the communication.

However, the effectiveness of a particular Communication process depends on a number of circumstances which include; the predisposition of the intended recipient; the basic needs of the individual; in what shape the message reached the recipient; the innate propensity to believe what is comforting to one's psyche or that shield it from guilt or fear; the skill and experience of the communicator. More still, we observe that a vast number of would-be communication is directed towards each individual every day. The skill of the public relations man then would be to select means and the context of the information he puts into the stream of communication, making it appeal to the recipient in his interest. This is because the effectiveness of the public relations function depends on how well information and ideas it communicates are adopted by the targeted individual or group.

Furthermore, some public relations masters do support the view that for communication to be easily adopted by a particular group, the immediate target audience should not be the entire group but the opinion leaders in that group. This point of view believes that once the opinion leaders buy an idea, that they would easily influence the general publics (Nwosu, 1990).

More still, communication must be based on factual data and information since these days, there is an increasing impatience with lies, half-truths and negative propaganda. As a result, patently insincere communication not only are ineffective but build resistance that will prevent acceptance of future efforts. Little wonder public relations is maintained to be based on truth and full information. We equally note that the previous reputation of the communicator aids in creating a "posture of receptibility" from the audience. For example, when effective public relations has been practiced for an organization and its actions and statements have developed positive image and goodwill, not only is a reservoir of support developed, but every other message from that source will receive much more acceptability.

Other salient points as regards communication which would aid the public relations person to achieve his communication objective include that the more closely a communication is beamed to specific audience, the more likely it is to be received and accepted. Also, the more sharply the key point of a communication is focused for the recipient, the more likely he is to grasp it. More still, it has been observed that establishing an idea in the public mind calls for a multiple combination of impressions which impinge on one's attention; the impression is created that the idea is all-pervading, that it is the thing to do". Hence it has considerably greater influence. Moreso, in designing communication material, care must be taken not to put art ahead of communication. It must be understood that the purpose of graphic materials in Public Relations is to aid the communication process. But when artists or designers emphasize their creativity or artistic capability at the expense of conveying the message, the result becomes a kind of "discommunication" or in communication. For example a graphic write-up that defies legibility repels the audience instead of communicating (Nwosu, 1996b).

The Education Component

Another important component of the PR RICEE model is education. Environmental education in this case is all about deep environmental knowledge or sensitization. According to the United Nations Environmental Programme report on the state of the Environment, there are three basic spheres of environmental education, viz; the home, the community and the school. They maintain that efforts in all these spheres must be linked and harmonized to create appropriate perceptions of environmental problems as well as solutions based on environmental awareness UN, 1990.

Integrated environmental education starts in the home and neighbourhood. Hence, girls and women are to be properly enlightened since they play very important role in this regard. Education at home plants ethics and seeds of future attitudes, which instills patterns of behavior that lead to marked healthy and sustainable environment. Mental alertness towards the natural environment seems to develop at the age of nine or ten. Children then can appreciate the interactions of people and nature (UNICEF, 1990). Thus, this presents a challenge for teachers, curricula designers, activity planners and educators. Hence to catch the people young and a lot of commitment is needed on the part of teachers of these age-groups.

In terms of formal education, some countries have introduced environmental education in primary schools. In others, it has been introduced as an added component of existing subjects: hygiene, nature study and population.

However, we note that there is an essential difference between 'learning' and "awareness". For example, a student may learn and understand that a particular plant is rare and may know a great deal about its geography, taxonomy, etc, but he may still pull it out by the roots. Hence true environmental appreciation means an awareness of nature's life-support, life-giving and aesthetics significance. For example, a child that is made aware of trees' protective functions or of the inherent beauty of flowers in their natural setting, will not uproot them.

This education would be a programme or enlightenment campaign package designed after appropriate research and information have been gathered and analyzed. Thus, the components of the PR RICEE model complimentarily play indispensable role in designing an effective educational enlightenment programme. For an environmental education to be relevant, it must be.

    (a) Value oriented

    (b) Community oriented and

    (c) Concerned with human well-being

Hence a successful environmental enlightenment campaign would have been achieved if the attitude and behaviours of a particular problem that takes place in a particular environment affects other areas of the globe, either in the short run or long run basis. The major purpose of such a campaign should, therefore, be to enable the masses or targeted groups or publics to understand the enormous evils of environmental degradation and pollution so that they can appreciate the urgent need to apply public relations and other environmental management in working hard to build and always maintain a sustainable environment in order to ensure balanced, safe and sustainable development all over the world.

The Evaluation Component

The last component of the RICEE model is the evaluation component. This is a very crucial part of any project such as environmental management project. The project life cycle model (Nwosu, 1990) stresses this point. Evaluation helps us to learn from hindsight in order to make necessary adjustments. It also helps us in planning or moving the project forward. Evaluation is always done at the end of the project. It is indeed a kind of research or systematic information or data gathering, this is why it is often described as evaluative research, as opposed to diagnostic research. When carried out along with on-project monitoring, it is very useful in ensuring high level of effectiveness in EPR and how badly we have done in the planning and execution of any project. So, it is an indispensable tool in environmental project management. So, the EPR manager should endeavour to master project evaluation techniques and strategies, as well as apply them correctly in any EPR management project he is involved in. this is the final and conclusive message of the RICEE model at this point.

Chapter four

The Environments of Environmental Public Relations (EPR) Management

Introduction

The title of this chapter may sound deceptively tautologous. It might even sound like an expensive and unnecessary pun or playing with the word, environment. This is understandable because in the study and practice of public relations itself we have until recently been only familiar with the standard practice of trends analysis, environmental scanning or systematized analysis of the environments of public relations practice – the internal competitive and external environments. And in this analysis, the ecosystem or the physical environment is merely analyzed as one part of the external environmental factors that the public relations manager must manage, like any other business manager.

But since EPR merged as a specialized and full-fledged area of public relations management or practice, as we explained or described it in chapter one of this book, we must identify, analyze and fully understand the ever-changing environmental factors that influence its practice so that we can properly manage them in the unique context of EPR practice and applications in various organizations (business or non-business). This is the task of this chapter.

For our purpose in this chapter we shall define the environments of EPR as the internal and external factors, actors and forces that are of interest to the EPR manager because they influence his organization's ability to successfully achieve the stated EPR and ecosystem objectives of the organization. They therefore need to be continually monitored, analyzed and managed to benefit from the ecosystem and EPR *opportunities* which they offer, with the best of the organization's strengths, deal masterly with the threats which they pose, and ensure that they do not successfully impinge upon or exploit the ecosystem-related and EPR *weaknesses* of the organization.

In formulating and adopting the above definition, we must point out that we are very much aware of the dominant or popular definition of the concept of the business environments, the marketing environments or public relations environments by various experts in different contexts, including even ourselves. For example, Philip Kotler (1994) the Marketing guru, has defined the marketing environment as "the external actors and forces that affect the company's ability to develop and maintain relationships with its target customers". Ikechukwu Nwosu (1996 and 2003) has also similarly identified and defined the environments of public relations management or practice, but replaced "target customers" with the "target publics" of the organization. But for a more indepth, unrestricted and thorough analysis and understanding of the environments of this new and specialized area of public relations practice, we have intentionally included the "internal factors, actors and forces" and veered off from restricting EPR to the organization's transactions with its external publics which includes customers. As can be seen from our above definitions of EPR, we have stressed on managing the influences of these factors, forces and actors on the organization's ability to deal successfully with ecosystem or environmental issues and problems. And to make this possible we are convinced that there must be a conscious effort by the organization through its EPR managers to identify, monitor and manage both the internal and *external* factors, forces and actors holistically with solid and well-thought-out EPR programmes of actions and communications.

All these are covered by what we have in different context described as ENVIRONMENTAL SCANNING. And we have always defined environmental scanning as:

> The process of systematically identifying, collecting, collating, analyzing and the packaging of the facts, problems, issues or data in any organization's *Internal* and *External* environments, then feeding all these into the management system of the organization to guide *managers decisions* and *actions* (Nwosu, 1996).

In the context of this chapter and this book, the managers referred to or being focused upon are EPR managers and other managers above, below or at the same level with them in the organization's hierarchy. And the decisions and actions referred to in the definition are those that are on or are related to managing the natural or physical environment or the ecosystem.

In all these, we should never forget that the key purpose of environmental scanning is essentially about understanding and managing the internal and external CHANGES that take place in an organization's internal and external environments with the aim of avoiding or dealing with the dangers, time bombs, minefields or problems of threats in these environments that may hinder the organization from achieving its set objectives (in this context, environmental management objectives), help it to maximally exploit the opportunities in these environments for achieving these set objectives, using the best of their already uncovered or discovered strengths and carefully managing the already identified weaknesses of the organization's environments in such a way that they do not thwart the organization's effort or get exploited by the opposition or "competitors" (in this context, the anti-environmental management actors, factors or forces).

It is indeed from all these that SWOT or TOWS Acronyms that have been in the management literature for a long time, was developed. In both acronyms, S stands for Strength, W stands for weaknesses, O stands for Opportunities and T stands for Threats. Both acronyms are quick but effective tools for analyzing the data collected by managers during the above-discussed environmental scanning exercise or effort. Whether we use the SWOT acronyms or the TOWs acronyms depends on such factors as the peculiar situations or problems which the organization is facing at any particular time, our prioritization of such factors or the problems and other such considerations.

With all these as background, we can now go ahead to identify and discuss the usual or general factors we should monitor and analyse from time to time, and from the internal and external environments.

MODUS OPERANDI

But before we proceed, we must point out further that for our purpose, we shall not strictly follow the dominant approach in analyzing the environments of an organization in marketing and management's studies or practice. That dominant practice is to classify or divide all the environmental factors into two-MICRO ENVIRONMENT and MACRO ENVIRONMENT and analyze the factors, forces or actors in each of them. We shall slightly deviate from this approach by dividing or classifying the environmental factors into three – the *internal environment*, the *micro environment* and the *macro environment.*

This slight but important modification or expansion will help us to capture the dynamics of the factors "within" or inside any organization that will go a long way to determine how able or unable that organization will be, in managing or handling or dealing with the changes which the influences from the factors, actors or forces within the other two environments, the micro environment and the macro environment which are actually external to the organization (the external environments), will have on its ability to achieve or not achieve its set objectives (in this context the ecosystem objectives). The micro-environment, of course, refers to those actors, forces or factors that influence and interact with any organization (from outside it or external to it) directly, close and regularly or continually (e.g. community actors, publics or customers, the mass media, interest groups competitors or opposition, financial or funding institutions or agencies, governments etc). And the macro environment refers to the wider or more general socio-economic and political domains including the ecosystem itself, whose factors, forces and actors influence the organization's internal and micro environments or the changes/developments within these internal and micro environments (e.g. economic, cultural, political, technological factors, etc).

The Internal Environment

The saying that charity begins at home applies here. Any organization must first keep its house in order before trying to get or reach out to manage the external factors, actors or forces in its external environments (the micro and macro environments). This is extremely important for any organization or agency that wants to get involved in the murky waters and highly sensitive issues and problems (or even people) involved in environmental or ecosystem management.

The first duty of EPR manager here therefore, is to work with other managers to ensure that the organizations internal environment's factors are available, adequate, strong and well managed. These internal environment's factors can be identified and described with what we have described as the 4 Ms + S formula (Nwosu, 2000). In the formula, the four Ms refer to the usual men, money, machine and materials in management studies or practices, while the S refers to Structures or physical structures (e.g. buildings, offices, spaces or landscapes, etc). Of all these five factors, the men or human resource is the most important in ecosystem or environmental management, and indeed all other areas or applications of management. The recruitment, enlistments or mobilization of the men and women that will be involved in managing environmental, "green" or ecosystem issues and problems must carefully or painstakingly be carried out by the organization or agency involved or concerned. They must be round pegs in round holes. They must be properly trained in EPR and general environmental issues, ecosystem management and general management strategies and techniques as well as properly motivated in a manner that goes far beyond money and Abraham Maslow's lower order needs (Maslow, 1954).

When the human resource element is adequately handled, it is much easier to manage the money or financial, machines or equipment, materials and structures dimensions of the internal environment or the organization or agency, including any changes or influences on them from the micro and macro environments. Money is another internal environment's factor that must be properly managed (e.g. sourced, controlled and carefully budgeted and well accounted for by the EPR manager and the rest of the management team. This is because environmental management activities (e.g. EPR public enlightenment and publicity campaigns, event management, community relations, crisis management, environmental impact analysis, researches, etc) can be very expensive indeed. We also need adequate amount of money to be able to procure the required machines or equipment, and materials, as well as set up adequate physical structures for carrying out result-oriented or effective EPR activities in any organization or agency.

Then, as part of managing the internal environment of his or her organization for effective ecosystem management, the EPR manger should work closely with his Chief Executive Officer (CEO) and other managers to ensure that a strategic management plan that is environmental friendly is put in place. This will help him to ensure that the vision, mission, corporate policies, corporate strategies, corporate culture and corporate objectives of the organization have written or unwritten pro-ecosystem dimensions. This will go a long way to make the EPR manager's job much easier. Both in terms of ensuring that his organization or agency is seen and known as one that has an environmental-friendly reputation and image, as well as in terms of EPR manager's task of designing, planning and executing specific EPR programmes and projects. Having finished with the internal environment of the organization, we can now move to the external environment, bearing in mind of course that for our purpose, we have subdivided the external environment above into the micro environment and macro environment.

The Micro Environment

We have already defined or stated above what we mean by the microenvironment in analyzing the environments of environmental public relations (EPR) and have given some examples of its constituents or elements. At these points, we shall discuss some of them for greater understanding and more in-depth treatment in this book.

The Governments

Governments are very important members or actors in any general environmental and EPR management situations. This is, understandably, so because of the most powerful or inexorable powers usually wielded by governments through their executive, legislative, judicial powers as well as policies and extra-legal pronouncements on environmental issues and problems, as in other areas of governance. The actions or inactions of government actors, especially in the developing countries with high level of poverty, ignorance, corruption, absences of human rights, poor governance, etc, (e.g. Nigeria) can spell doom or success for any general environmental or EPR programme or project. It does not matter if the government involved in the programmes or projects is the central government, state government or local government. It could even be just a government parastatals or agency like the Federal Environmental Protection Agency (FEPA) of Nigeria or the state government environmental agencies in Nigeria's 36 states and Federal Capital Territory (FCT) at Abuja, that will have those negative influences through their inactions or actions. Even when they take actions, on environmentally related issues or problems, these actions should be above par or at least adequate, for them to have the desired impact. The problem though is that most of the time these governments are guilty of inaction or inadequate actions that have continued to worsen the environmental issues and problems in these developing areas.

Yet, the EPR manager does not have a choice. He must keep working hard always to find a way of enlisting the support and understanding of these government agencies. If not for anything, the EPR manager needs substantial and regular funding from these government or government agencies to finance most of its environmental programmes or projects, many of which we know can be very expensive .The EPR manager does not have any choice most of the time especially in the developing countries, because these governments control most of the fund or money available in the country, state or local government concerned. He must therefore find a way of penetrating these governments and government agencies, using various tried and tested public relations strategies (e.g. lobbying), because he needs them badly as partners in progress, as far as environmental management is concerned.

Apart from funding, it is only when these government and government parastatals see themselves as partners in progress with the EPR and other environmental experts that they can come up with environmentally friendly legislations and policies as well as be ready to work closely with environmental experts to see that these are implemented and widely adopted by the citizenry.

For sure, since the 1992 RIO Summit, environment protection has been a major concern and issue in the agenda of governments all over the world. What is needed however, is not just expressions of concern by governments but concrete actions. And in this respect the developed countries' government are very far ahead of the developing countries' governments. Unfortunately, it is in these developing countries that we have the highest need for environmental protection since they have the worst cases of environmental pollution or degradation whose effect on the citizenry are made worse by poverty. So, the governments of the developing countries like Nigeria do not just have to wake up and improve significantly their environmental management practices, but in the spirit of the global village theory or globalization trends with the slogan of one world, one destiny, the governments of the developed countries should, on the platform of appropriate global agencies(eg. UN Agencies), team up with the governments of the developing countries in the war against environmental degradation. Here again, the EPR manager and other environmental experts have a key role to play in advocating, lobbying for and promoting this global approach for environmental management through greater and more integrated involvement by the governments of the developed and developing countries.

As we noted earlier, we cannot do without the government in managing environmental issues and problems, especially in the developing countries. We have tried to give various reasons for this situation above. But similarly we must additionally note that government policies tend to contain an admixture of command and-control style regulations. Financial incentives and the promotion of industry self-regulation. And these can affect the prices of environmental products, quantitative aspects of eco-performance such as outputs of emissions or the nature of the technology used (Simon, 1992:268-285).

Host Communities and Community Neighbours

We cannot neglect local communities in any effort to manage environmental problems and issues. For one thing they feel the negative impact of environmental pollution and degradations in many ways, especially in the developing countries where about 75-80% of the citizenry or population still live in the rural areas, where poverty is most severe. Next to them in the "environmental suffering ladder" are the residents of the urban slums.

The EPR manager should therefore, put his community relations and corporate social responsibility (CSR) strategies to the maximum use in managing environmental issues and problems at the local community levels. This will help them to properly mobilize the members of the host communities of their organizations, as well as their community neighbour to ensure the highest level of participation and partnership with them from the beginning of any environmental promotion or protection project to the end. To this end, the EPR managers should ensure to enlist the support of the town unions or development unions (e.g. the Amawbia Town Union ATU), community associations like age-grades for a (e.g. Udoka Age Grade of Amawbia), community leaders and chiefs (e.g. the Okpaligwe of Amawbia) the women groups like the "Umuada and Umu-okpu" and "Awo Mmili" associations, all in Amawbia town in South-East, Nigeria.

These groups or actors or others like them in most local communities are very influential and do not only influence the attitudes, opinions or behaviours of the citizenry, but always have, well established methods of group or individual sanctions that make for high compliance by most of the citizens, if not all. The local groups and citizens wear the environmental degradation shoe and so usually know where it pinches most. We cannot do without them. We must work with them. And for greater effectiveness we must understand, the concepts and influence of community structures, values, mores, norms and culture in general. You cannot work with or respect people you do not understand their ways and they cannot, of course, respect you or your organization, not to talk or working with you to find solutions to environmental issues and problems. The host community refers to the immediate community in which your organization exists and operates (e.g. Awka Town, Nibo Town, Nwafia Town, Ugwuoba Town, Enugu-Agidi and Nise Town, all near Amawbia Town (Nwosu, 1996).

Banks, Insurance and other Financial Institutions

"No fund, No EPR" is a maxim that should always guide the EPR manager, if he wants to achieve high degree of success always for his projects, as he should. EPR and other environmental management activities can be very expensive indeed. The actions and inactions of banks, insurance and other financial institutions in any society or country, therefore, have serious influence, on the environmental management quality of that society or country. These financial institutions are also influenced by the environmental problems, issues, and regulations and policies in the countries and societies where they operate.

It is therefore, imperative that the EPR management of these financial organizations (e.g. bank and insurance companies) work closely, with EPR consultants, experts and managers outside their own companies as actors in the ecosystem (e.g. governments), to formulate policies, initiate policies or ensure the implementation of environmental protection policies and regulations in their areas of operation and beyond. They must embrace the "Green Culture" by making all their policies, products or services and actions or activities to be environmentally friendly in their own interest and in the interest of others in their immediate and remote environments. The achievement of this objective holistically is one of the biggest challenges of the contemporary EPR manager. He must among other things be an expert in financial public relations.

Investors and EPR

Investors, like banks and other financial institutions we discussed above, are major actors in the environment of environmental public relations (EPR). The EPR management should therefore not forget to sharpen his professional tools or strategies in the broad area of investor relations and apply them systematically in his EPR projects to relate to the highly competitive and project driven business sector. To this end, he and his team should never get tired in reminding this business investors of their roles and social responsibilities towards managing the ecosystem or physical environments in which their businesses operate. They should also educate and keep reminding them of the consequences of their neglect of these environmental roles and responsibilities both for the country and the world.

Through these and other EPR methods, the EPR manager and his team should be able to enlist these investors and their businesses or companies in the war against environmental degradation. In doing this, the EPR manager and his team may need to enlist the support of various groups in the organized private sector (OPS). Such groups in Nigeria include the Nigerian Association of Chambers of Commerce, Industry, Mines and Agriculture (NACCIMA) at the centre, the Enugu Chamber of Commerce, Industry, Mines and Agriculture (ECCIMA) in Enugu South Eastern Nigeria, The Lagos Chamber of Commerce, Industry, Mining and Agriculture in South Western Nigeria (Lagos), and their sister body in Northern Nigeria, as well as the central/national and state branches of the Manufacturers Association of Nigeria (MAN).

Unfortunately, the EPR managers in Nigeria and other developing countries would have to work harder than their counterparts in the developed countries because facts available in the environmental research literature show clearly that the awareness on and acceptance of environmental problems and responsibilities are already high or more established in these developed countries. One of such studies carried out as far back as 1991 shows that in most European countries, there is a rising interest in green investment and a recognition that a greener strategy can generate competitive advantage and cost efficiency (Tennant and Companale, 1991). Another survey research carried out the same year by Dewe Rogerson (1991) in which he studied 80 senior fund managers form major institutional investors, found among other things that:

- 67% of the investors/managers believed that environmental factors had a significant effect on business;

- 58% said that a coherent and effective environmental strategy enhanced their perception of a business;
- 90% said that environmental performance was an important factor in their investment decisions;
- Only 11% viewed the environment as unimportant.

It would be very interesting to survey Nigerian or Zambian investors to find out if they would be this much driven by environmental considerations in investment decisions, policies and actions. The findings, compared to the figures in the above reported study, may be very revealing negatively, if not shocking. Again this translates to a big challenge for the EPR manager operating in these and other developing countries who would have to start at the basic level of sufficient awareness creation before daring into the higher levels of attitudinal opinion and behavioural changes among the investors in these countries, in order to enlist them as true partners in progress in any effort to manage their physical environments or ecosystem problems more effectively.

Media and Media Practitioners

Media relations' management is a crucial part of professional public relations practice and management. It involves, among other things, a deep understanding by public relations managers of the mass media as a social institution and as a business enterprise as well as understanding media practitioners (editors, reporters etc), their professional needs, wants and interests and how all these interact with the other intra-media and extra-media dynamics to determine whatever role the media plays in any sector (local, state, national or global), whatever effects the media have and how the media itself is influenced by forces external to it (Nwosu, 1996; Nwosu, 2000). The EPR manager should not only know all these but should know how to factor the media and media men into his EPR management equation. Media and media men must be persuaded or won over by the EPR manager to partner with him in his effort to achieve EPR objectives. It will be a very formidable partnership as the facts and figures in the role of the media in the management of environmental problems and issues, which we have critically and quantitatively handled, or examined in various chapters of this book, have shown.

Furthermore, with the growing influence of the satellite and ICT-based media, the media influence in managing ecosystem problems will be even more pronounced. Also, media practitioners who by their training, often see themselves as the watchdog and conscience of the society as well as the eyes and ears of the masses (Nwosu, 1990), now tend to see themselves as crusaders or advocacy instruments against industrial and government agencies, companies or other groups or individuals who engage in any activities or formulate as well as implement policies that are inimical to environmental protection or preservation. Many of them have on their own initiated pro-ecosystem or green campaigns, using such media fares as editorials, commentaries, articles, features, advertorials and others.

Ken Paettie (1995:61-63) would agree with the above observation and has written that "The environment has become increasingly visible as a media issue... the more dramatic and therefore newsworthy environmental degradation becomes the closer the interest that media organizations will take in corporate environmentally related deeds and words". Has this practice or trend always been like this and in all parts of the world? The answer is Yes and NO because the media dynamics we referred to earlier in this chapter have always ensured that there will be some deviations from the norm or trend from time to time and from place to place around the globe. The different studies of mass media handling of environmental issues and problems, which we reported in various chapters of this book, clearly reflect these ever-changing trends.

The studies carried out in Britain by different researchers on media reportage of environmental issues and problems also confirm this ever-changing trend over time. The study by S.K. Brooks and his research associates (1976:245-255) reported that in newspaper journalism there was steady but minor coverage of the environment in the pages of *The Time* between 1953 and 1965. This was followed by triple or three-fold increase in this coverage between 1995 and 1973 in terms of proportion of space devoted to environmental issues. And A. Mitchel and L. Levy (1989) reported that in Britain between 1985 and 1989, the use of the phrase "environmentally friendly" in a sample of printed media leapt from once a month to 30 times a day, and the word recycling was used nine times as often".

All said, the point must be made that due to the increasing attention being paid worldwide to the environment, the greater awareness on the environment in different societies and improved ICT developments, environmental issues are now (1990s and early 2000s) generally in the increase and there are also increased media advocacy or adversarial campaigns on the environment, with the media in the developed world understandably leading or doing much better than those in the developing countries. According to D. Mulhall (1973). "the massive impact of instant media in accelerating the message of gross environmental incompetence by our leaders can be summarized in three letters – CNN. It means that a company's reputation can be destroyed globally in one day".

Green NGOs, Interest and Activist Groups

The last group, but by no way the least group, that the EPR manager should pay serious attention to, is made up of many environmentally relevant groups we shall lump together for the purpose of discussion in this chapter, but which are really many and usually autonomous. These are the Non-Governmental Organizations that focus their interest on the environment (the green NGOs), as well as other special interest and activist groups who are very concerned and involved in drawing attention to environmental issues and problems, combating anti-green behaviours or policies by governmental companies or individuals and taking other direct and indirect actions to help ensure a safe, clean and sustainable environment in any society. Apart from working independently, these groups are very good in partnering with other groups or individuals to achieve their environmental protection objectives. They are also perceived by the general public as credible, effective and altruistic in their policies and actions.

These are parts of the reasons why the EPR manager working within an organization must know them, understand them, listen to them and work closely with them, to ensure that their organization's environmental objectives are well set, remain on track and get achieved as well as become seen by the general public (perception) as having been achieved and can be sustained. They should work closely with them to avert negative perceptions of their organization's policies or environmental mistakes by their organizations that can attract serious consequences from the general public, the media, the government and other stakeholders. EPR consultants should also work along the lines we described above to ensure success all the time in the environmental management policies, objectives and activities of companies or organizations they represent.

EPR managers should also be very familiar with the popular methodologies of the pro-environmental interest groups or how they operate, as well as the key people, representatives or officials in these groups. They should move from them to build public relations bridges of mutual recognition, mutual acceptance, mutual trust and mutual respect, because this is the only way they can work well with them all the time or on specific environmental management project.

Their major methods of operations, in addition to the ones we mentioned above include *direct Action* emphasized by such groups as the Green Peace Movement, *Campaigning and Lobbying* emphasized by groups like the Friends of the Earth, and *Partnership* emphasized by such groups as the World Wide Fund for Nature (WWF) - - all depending on the situation or circumstances being confronted by the group concerned. It is also important to note that whenever an organization is under attack from these pro-environmental groups, the EPR manager is advised not to adopt the usually negative responses like Defence, Denial, Discrediting, Disapproval and Deflection. We like to call these the 5Ds of negative response to environmental criticism, attacks or complaints against your organization.

The EPR manager is advised to avoid them like AIDS or leprosy and SARS because as K. Peattie (1995:64) rightly pointed out, "Such responses can be counter-productive, since given conflicting stories from green groups and a company, the company is likely to finish second in the race for credibility". Peattie supported his point with data from a study by W.K. Kamena (1991) which surveyed selected consumers in the U.S.A. to find out how they score or rank-order different groups as sources of environmental information, and found out that "environmental groups scored 37%, the highest score, consumer groups 24%, the media 19%, Retailers 9%, Government 7% and product manufacturers a very distant 5%. So, are you dying or longing to fight with the environmental interest or activist groups for credibility? We guess you would rather not.

Competitors and Customers

We have intentionally delayed the discussion of such hard-core business or cash profit-oriented elements of the micro-environment of any company like competitors, customers and marketing intermediaries, like the distribution channel, members (e.g. distributors and retailer), suppliers and price or pricing. This is because this book is dealing with a not-for-profit issue – the environment. But since these elements are still important conceptually and operationally in not-for-profit or social marketing, we shall discuss some of them only summarily, as part of the micro-environment of any organization involved in managing environmental issues and problems. Moreover, this is a book on Public Relations, not Marketing.

With regard to competitors, the job of the EPR manager is to regularly monitor what competing organizations are doing in the area of environmental management, then design and implement systematically a programme of actions and communications that will help to ensure that his organization maintains the competitive advantage in terms of corporate performance in the area of environmental protection. He should also advise his management regularly on what to do or not to do in managing environmental issues and problems within the company and outside the company, including contributing from time to time to environmental protection causes or efforts, as part of the organization's corporate social responsibility programme. The EPR manager and his team should remain proactive rather than reactive when dealing with environmental issues and problems.

They should also ensure that the *packaging* of their organization's products are environmentally friendly and environmental laws complaint.

In respect of pricing, PR managers should advice continually that their organizations, especially those who produce and sell products or services for profit, differentiate between *monetary price* from *socio-environmental price,* which is not aimed at immediate cash profit, so that they do not regard what they spent on the environment as economic waste. They should also plan and execute campaigns aimed at educating and persuading consumers and companies to purchase less environmentally damaging or more environmentally friendly products and services, even at higher prices. Even more importantly, we believe that EPR manager should continue reminding all stakeholders in any society that the environment or nature is indeed priceless. It is thereof absurd to consider whatever inputs or investment we make in preserving the environment in terms of the price they cost us in terms of money or even time, which, we know, is money.

Peattie (1995:239) will probably agree fully with the above thoughts process and has captured the dilemma or difficulty in considering price in environmental management in these words:

> Societies and consumers pay two prices for goods and services that are consumed. The economic price is clearly defined, must be paid today or soon after, and any debt that

is incurred can be paid later. The socio-environmental price is *unclear*, can be deferred today (even if it must be paid eventually) and any debts incurred are unlikely to be paid.

As far as the suppliers and marketing intermediaries are concerned, all we want to say here is that the EPR manager and his team should regularly check the environmental records or performance of these marketing actors and advise management whether the company should continue to do business with them or not.

What about the customer as part of micro-environment in EPR management? The main job of the EPR manager and his team here is to regularly monitor and control or manage, without panicking, customer or consumer complaints and pressures that are related to the environmental performance of their organization and its products or services. They should also use preventive EPR strategies to avoid such complaints or pressures. They should in addition, carry out research and maintain data banks of past, present and emerging customers' of consumers' of environmental or ecosystem concerns, especially as they relate to their organizations, in order to address these concerns pointedly, appropriately and adequately. They should also, as recommended by the Total Quality Management (TQM) strategy, mobilize and carry the internal customers (employees) along in all EPR campaigns.

The Macro Environment of EPR

The PEST acronym or analytical model has been for a long time the most popular tool for analyzing the macro environments part of the external environment, as well as the influences they have on the organization and its activities. This method of the macro environments analysis is sometimes simply known as carrying out a PEST Analysis. P stands for *political environment* and usually includes the *legal and regulatory* environments. E stands for *Economic environment*, including national and global economic trends, changes, cycles of recession, population trends etc. S stands for *social environment,* which includes changes in values, attitudes, opinions, lifestyles, ethics and other socio-graphic variables. T stands for the *Technological environment,* which includes technological changes like the Internet, Database, Digital Television and many others (Smith, 2000:188; Peattie, 1995:65).

We recommend however, that the EPR manager and his team go beyond this basic analytical tool (Pest Analysis) in any attempt by them to analyse the trends or changes in the macro-environment of their organizations (environmental scanning) to gather data and information that will guide their ecosystem decisions and actions, as well as those of other managers in the organization. Why? For one, the PEST model is somewhat limited in its scope and analytical tool, despite the attempts of those who use it to subsume related variables under any of the four sub-sectors, or themes that are closely related to them. For example, as we noted above, the legal and regulatory environments are subsumed or forced to fit into the political environment. But, as we should know, such subsumed treatment of any issue or factor or variable does not make for indepth or sufficient analysis of all dimensions of such issues. Moreover, the subsumed factors or issues are crucial or important enough to be given their own independent analysis or considerations. Again, in any attempt to put everything or every factor in the PEST bag, some very important issues are most often totally forgotten. Therefore, the EPR managers should use environmental analytical models that are more broad-based and which will help them to do proper and in-depth environmental trends analysis for both profit or commercial and non-profit of non-commercial organizations, and which gives them the opportunity to carry out very deep analysis of the physical environment or ecosystem which is the major focus of EPR.

To this end, we recommend that EPR managers adopt and apply (as it is or in any contextually modified form) Nwosu's SCEPPTICALE MODEL or environmental trends analytical framework (2003:7), which has been used in environmental analysis of business and non-business organizations. This model was developed from the SCEPTICAL analytical model which has also been in the literature (Peattie, 1995) by the addition of two new factors or variables which, especially for developing countries, were considered as very important, but which have been neglected, subsumed or left out completely in environmental scanning efforts of managers and experts, especially those from the developed countries and the old business management disciplines, because they can afford to forget about them, take them for granted, pay brief or tangential attention or lip service to them in their analyses. One of the big but often forgotten variables is EDUCATION. The second variable highlighted in Nwosu's SCEPPTICALE model is POLITICS, which was surprisingly subsumed under the Legal environment in the SCEPTICAL model. It is erroneously put last in that analytical model. We consider this as strange because we all know the all-powerful and over-bearing influences of the political actors, forces and issues in all sphere of life, especially in the developing countries.

We had to add and make EDUCATION prominent in the new SCEPPTICALE Model because managers in the developing countries cannot afford to leave out or play down education or changes and influences from the education sector because it is the root of all types of development. And in most developing countries like Nigeria, the education sector is either wobbling, grossly neglected or sick. It is therefore, a major source of crisis or other problems that can negatively influence, if not destroy any plans or objectives we set for our organizations – short-term, medium-term, or long-term, including eco-system objectives.

Furthermore, education of the populace is central to the achievement of ecosystem or environmental management efforts or campaigns. In fact, it is always there in every EPR or general environmental management models or plans of action (e.g. the RICEE model). It does not matter whether we are thinking of formal, nomadic or any other types or forms of education; the changes, forces or actors in them will always play a major role or have major influences on ecosystem management, especially in the developing countries where functional illiteracy is still not only a general problem but also an ecosystem problem.

In addition, apart from working for or with orthodox or pure business organizations who may not see any direct connection between the organization's bottom line or profit (or even as a top line essential), EPR managers will definitely also work for or with not-for-profit organizations like UN agencies, NGOs, or even religious organizations whose corporate objectives may be solely or partially ecosystem management. They might even work inside educational institutions at the primary, secondary, or tertiary levels. So, the education sector or the educational environment is a sector or environment which EPR managers and other managers cannot afford to neglect in their environmental scanning efforts, both in their own interest and in the interest of the wider society in which they exist and operate. Indeed, it is corporate immorality to do so, especially in developing countries like Africa. The case for isolating and making Politics a separate environment in the SCEPPTICALE model is quite obvious and so requires not further explanation here.

In the SCEPPTICALE Model that we are recommending here to the EPR managers as an effective environmental analytical framework:

- S stands for the Social environment;
- C stands for the Cultural environment;
- E stands for the Economic environment;
- P stands for the Political environment;
- P stands for the Physical environment (the ecosystem);
- T stands for the Technological and Scientific environment;
- I stands for the International or global environment;
- C stands for the Communications environment (processes and structures);
- A stands for the Administrative, managerial and institutional environment.

- L stands for the Legal and regulatory environment;
- E stands for the Education environment.

Note that in the above analytical model or framework, cultural environment has been separated from the social environment to make for greater or more indepth analysis of culture, which is a critical factor in ecosystem or environmental management. Most experts lump them together and call it the socio-cultural environment. In EPR environmental scanning, this separation will allow us to do detailed analysis of the two vital factors in the environment or the ecosystem. Note also the prominent fifth position given by the model to the physical environment or the ecosystem. Whether we believe it or not, if we continue polluting, neglecting, raping and reaping recklessly from mother nature or mother earth, she will die a natural death, a slow, tortuous but sure death. And when she dies all the other factors in the model will not matter anymore. Why? Because dead men cannot set up organizations, not to talk of analyzing their environments and running or operating them.

But even if dead men can set up organizations, there will be no mother earth on which those organizations can exist. It is not yet conclusive that there is life, as we know it, in the other planets. But even if there is life in the other planets, we are not sure they will welcome human beings who have notoriously and without qualms destroyed or "murdered" their own mother-earth, habitat or planet. Even if they are so benevolent to welcome us, we are not sure that we will not continue with our bad old ways of destroying whatever good that is given to us by God. And if we are not sure of ourselves, how can they be sure of us. So, now is the time to change our anti-ecosystem behaviours because there is nowhere to run to after destroying mother earth. This should be the key message of EPR managers and all others interested in the conservation of our God-given environment or ecosystem.

The other elements of our SCEPPTICALE Model are easy to understand and apply. They are also interrelated and should be treated holistically, along with the internal and microenvironment factors we discussed earlier in this chapter. But we should take particular note of the communications environment in the model because of its great importance to the job of the EPR manager, the existence of any organization, community, country, society and indeed the global or international system which is also highlighted in this model because, as we all should know, the world is now a global village in which no one can live in isolation. And communication holds the key to interactions and survival in the new global village (mother earth), which we should all work hard to protect or preserve for our generation and generations unborn.

Conclusion

We like to simply conclude this chapter by reminding us all that there can be no sustainable development without sustainable environmental development or a sustainable environment. It is therefore, absurd to hear or read our national and world leaders and even some experts, discuss the theory and practice of sustainable development, or even development alone, without any mention of the environment or sustainable environmental development. This chapter which focused on how to analyze the environments of environmental or ecosystem management (internal, micro and macro environments), should provide enough insights or ideas that can be used not just by EPR managers but any other individual or group involved in sustainable development, sustainable environmental development and any form of development.

PART THREE

****

INDEPTH ANALYSIS OF
ENVIRONMENTAL PROBLEMS
AND ISSUES

****

Chapter Five
Vital Facts About Environmental Management

Introduction

One of the great tragedies facing environmental management, especially in the developing countries is ignorance of the issues and problems of the environment. The ignorance of the masses which we have described or lamented about earlier and which the EPR managers should address, is serious indeed. But it is even more serious when some so-called experts on the environment, including environmental public relations (EPR) experts, are ignorant of salient facts they should know about the environment and environmental management. That would be double jeopardy and terribly disastrous for man and the environment because, as the saying goes, the blind cannot lead his fellow blind. And you must first educate yourself before you can educate others.

This is why we shall devote this chapter to the specific clarification, discussion or analysis of selected environmental facts, information and data we consider vital in environmental public relations management. For sure, every chapter of this book contains many of such facts, information or data on the environment. But in this chapter we need to present some of them more pointedly and offer some others that were not specifically or directly mentioned in these other chapters.

On a general note, we want to start by restating very pointedly that EPR managers should know that TODAY most less developed countries (LDCs), including Nigeria, face very significant environmental problems topmost of which are inadequate sanitation facilities, air and water pollution, land degradation, deforestation, loss of biodiversify, desertification, drought, gulley erosion and others. But we must specifically note that most of these environmental problems have been shown by research to be closely related to poverty, population pressure, poor governance, as well as market and policy failure that include inadequate institutional structures for natural resources management, poor environmental planning, control and policy/project implementations, environmental neglect, poor capacity building, constitutional, national integration and such problematic factors. Social and political instability has also contributed to environmental degradation in most LDCs (United Nations, 1998:37).

For example, it is estimated that the African countries (e.g. Cote d'Ivoire, Sierra Leone) have lost about 80% of their forests before the close of the last century. And about 6,500 hectares of the African savanna are cleared routinely, and the effects of bush fires are increasing. All these other factors, for sure, conspire to continue threatening Africa's biodiversity, natural resources, food and natural habitat. Add all these to such negative phenomena as animal poaching or the so-called endangered species in Africa's rare animal species, which are increasingly becoming extinct. The population of the African white elephant had, for example, declined by about 50% by the 1980s. Just imagine what its population or rate of extinction will be today. The precarious state or the environmental problems facing the African continent can best be imagined when we add such facts as dumping of toxic waste; misuse of agro-chemicals and fertilizers; mineral exploration, rural-urban migration; improper reclamation of land and other such "ecological atrocities" (Onumonu, 1996:32-37; Ukpong, 1994).

To all the environmental problems we identified above must be added the issue of the ozone layer depletion, global warming and sharp/adverse climate changes which many Africans may feel are not yet a big problem in the continent, but are already here with us; even if they are for a moment, less pronounced in Africa as they are in other regions of the global/community which Africa is very much a part of. We must also add to the above environmental issues and problems that should concern us in Africa, the following issues which in 1996 were described as "new issues in the domain of environmental protection" by F.R. Touryatunga (1986:25-26), because for one thing, they are no longer really new and so are much here with us:

a. Sustainable development and the environment;

b. Environmental impact analysis;

c. Debt-for-nature swaps including eco-diplomacy, or bio-diplomacy (e.g. the World Wide Fund for Nature has helped Madagascar to write off $3 million of its debts in return for safeguarding some rare plants);

d. The polluter-pay-principles (e.g. you pollute and you must clean your mess policy ... Company, government or groups);

e. The Human Right/Commission's application of its charter of the right of everyone to "a fully meaning life;

f. The concept of one world; one global village; one global family.

With respect to the Nigerian nation, Ukpong (1994) specially identified the causes of land degradation to be improper resource management, destructive logging of our forests, overgrazing and over-cropping of arable lands, flooding and wind erosion menace, strip mining in Jos and Enugu areas, land degradation with pesticides and fertilizers, some known natural landslides and destruction of wetlands and marshes for developmental activities and oil drilling. Defining biodiversity as the existence of life and its processes (in line with the *Texas Environmental Magazine*), Ukpong stated that its loss is mainly caused by the destruction of inhabitants as lands are cleared for agricultural and development purposes.

According to him, the greatest loss is the disappearance of the world's tropical forests which house 40% of plants and animal species, especially in the oil producing areas like the Rivers, Cross Rivers, Delta, Akwa Ibom and Abia States of Nigeria where not only land but varieties of plants and animals have been and are still being destroyed by these development activities.

On the deforestation problem in Nigeria, the Food and Agriculture Organization (FAO), an agency of the United Nations, reported that the country through circles of exploitation and husbandry, destroys reforestation efforts of about 25,000 hectares every year and replenishes only 4% of this loss yearly. Gulley or soil and coastal erosion is another very serious ecological or environmental problem in Nigeria that must be given specific mention and specific attention because it has destroyed many living, farming and economic land areas in various part of the country, especially the south-eastern states. For example, Anambra State which is leading the entire Nigeria on this negative developmental problem by having as many as 150 gulley erosion sites, Cross River which has 130 gulley erosion sites and Akwa Ibom which has 59 of such sites (Okorie, 1992). This has increased at the time of writing this book (2004).

Yet, nothing really serious has been done by the Nigerian Government or its Federal Environmental Protection Agency (FEPA). Ditto for the UN and other global agencies involved in environmental protection or management. Of course, especially in urban Nigeria, the problems of environmental sanitation, disposal of management of solid waste, air pollution, water pollution, noise pollution, industrial and commercial pollutions, gas flaring and others are still so intractable that the dangers they pose to urban dwellers and humanity in general, are quite enormous. The big problem of desert encroachment or desertification and drought in North-Eastern and North-Western areas of Nigeria and the other environmental or ecological problems touched upon in other chapters of this book are all crying for decisive actions by relevant agencies and by you and I.

Outstanding Environmental Laws, Conventions and Protocols

The EPR manager should be familiar with the outstanding laws, conventions and protocols that have been instituted at international and national levels as interventions aimed at checking environmental degradation. There are indeed now so many of them in existence today. However, we must be quick to observe that most of these laws, conventions and protocols have remained toothless bulldogs; or at the best moral suasions that are supposed to appeal to national, global, group and individual consciences on the question of environmental protection or conservation. There has either been general lack of the will to seriously implement them or those that would implement them are too busy with other issues (e.g. politics and the economy) to pay adequate attention to implementing them. This is absurd and a grave misplacement of priorities because as various studies have shown and as this book will show in its many parts, environmental problems and issues are intimately intertwined with political, economic and social issues. Be it as it may, it is necessary for EPR managers and practitioners or consultants to know that it is part of their tasks to enforce these laws, conventions and protocols onto the public agenda using various EPR strategies, the media and other communication and persuasion/advocacy methods. Among the other global conventions and protocols to which many countries, including Nigeria, are signatories to are:

    i. The 1987 Montreal Protocol on Substances that deplete the ozone layer.

    ii.   The 1989 Base/Convention on the Control of Transboundary

Movement of Hazardous Waste and their Disposal and the 1992 Framework Convention on Climate Change. Others are:

    i. The 1992 Convention on Biological diversity;

    ii.   The 1994 Convention on Control of Desertification;

    iii.   The 1973 convention on International Trade on Endangered Species (CITES);

iv.   The 1972 Convention on the Prevention of Marine Pollution by Dumping of Waste;

v.   The United Nations Environmental Programme (UNEP); Food and Agriculture Organization (FAO) prior Information Consent (PIC);

vi.   The International Code of Conduct on International Trade in Potentially Toxic Chemicals and pesticides;

vii.   And others

The leading global and national laws (Acts, bye-laws, Decrees, rules, regulations etc), which the EPR manager should be familiar with include:

1.   ***The U.S.A. National Environmental Policy Act***

    (NEPA) instituted by the U.S. Environmental Policy Organization and the U.S. Congress following various ecological disasters including large-scale of spillages, that seriously damaged aquatic and marine life; instituted in 1996 but taking effect from January 1, 1990. Its objective includes using NEPA to carry out periodic Environment Impact Assessment (EIA).

2.   ***The Nigerian Decree 42 of 1988 on Harmful waste (Special Criminal Provision)***

    The decree which is now an Act of the Federal Republic of Nigeria makes it a criminal offence for any company or individual to engage in activities relating to the purchase, importation, transit, transportation, deposit or storage of harmful wastes in Nigeria (Aina, 1990:76).

3.   ***The Federal Environmental Protection Agency Decree 59 of 1959***

This is the decree or law that formerly set up Nigeria's (FEPA), which has similar roles as the USA's (NEPA)discussed above. It is the major body empowered to regulate and control environmental issues, including ensuring air quality, water quality, atmospheric and ozone protection, noise control and control of hazardous substances (Aina, 1990:76).

4. ***The Environmental Protection Edict Number 13 of 1989 in Lagos State***

This Edict is similar in function, provision, and scope to the laws described in Numbers 2 and 3 above, but its powers are restricted to the Lagos State. It is however, more specific on the prohibition of such activities as digging wells or boreholes for industrial purposes without permission; and burning or causing to be burned such things as refuse of any kind, bushes, weeds, grass, tyres, cables or wastes of any description or use of tyres as source of fuels, without first obtaining written permission from the Ministry of Environment and Physical Planning of Lagos State (Aina, 19902: 78-79).

5. ***The Environmental Impact Assessment Decree Number 86 of 1992***

This Nigerian decree or law is supposed to be enforced by FEPA after proper EIA has been carried out. Among other things this law stipulates that no project will be embarked upon or authorized by the public or private sector of the economy in Nigeria of their environmental effects or implications (Umeh and Uchegbu, 1997:17-20).

6. *Other Environmental Protection Laws or Regulations in Nigeria*

Related to the above major laws and complementing them and each other in many ways are the following other laws and regulations:-

(a) The urban and Regional Planning Decree Number 88 of 1992, relating to setting up of buildings and such structures in Nigerian urban towns.

(b) The National Policy on the Environment, which emphasizes sustainable environment and sustainable development (see Appendices).

(c) The 1991 National Guidelines and Standards for Environmental Pollution Control in Nigeria, a basic regulatory instrument for monitoring and checking industrial and urban pollution.

(d) The National Effluent Limitation regulation, which is Section 1.8 of the above cited mandatory requirement for industrial facilities to install anti-pollution equipment in their organizations, as well as prescribes maximum limits of effluent parameters allowed for discharges.

(e) Section 1.9 of the same National Guidelines and Standards ... (cited above) contains the pollution Abatement in Industries and Facilities Generating regulations, which among other things provides restriction for the release of toxic substances, strategies for waste reduction and the safety of workers.

(f) The Sectoral Guidelines for Environmental Impact Analysis (EIA), which stipulates the detailed guidelines for conducting environmental impact assessments for projects on sectorial basis.

(g) The natural resources Conservation Plan which prescribes sectorial strategies for the conservation of Nigeria's National resources (Fauna, Flora, Soil and Aquatic ecosystems) towards sustainable development (Adegoke, 1997:5-7).

(h) The Lagos State Environmental Sanitation Edict Number 12 of 1985 as amended by Edict Number 4 of 1987

(i) The River Basins Development Authority Act of 1977

(j) Mineral Oil (Safety) Regulatory of 1963

(k) The Petroleum Regulation of 1967

(l) The Oil in Navigable Waters Act of 1968

(m)   The Oil Terminals Dues Act of 1969

(n) The Oil Pipeline Act of 1956 as amended in 1965

(o) The Petroleum (Drilling and Production) Regulation of 1969

(p) The Petroleum (Refining) Regulations of 1974

(q) The Petroleum Production and Distribution (Anti-Sabotage Act of 1975).

(r) The Associated Gas Re-injection Act 1979.

(s) The Criminal Code Cap. 42, volume 2, Laws of the Federation of Nigeria in 1958, as 245-247.

ENVIRONMENTAL AGENCIES IN NIGERIA

There are four major types of environmental agencies recognized and empowered in Nigeria to monitor, regulate and control various environmental degradation or destruction issues, problems and activities in Nigeria. These are made up of the United Nations Organization Agency (UNEP), FEPA and other state agencies discussed above, voluntary Environmental Advocacy Groups, and the Community-based Environmental Protection Programmes. Let us briefly discuss them.

The United Nations Environmental Programme (UNEP)

UNEP was set up in 1972 by the UNO as the apogee governing council of environmental programmes and projects in the world. It has its headquarters in Nairobi, Kenya and actually started operations in 1980 providing support for research, programmes, projects and other environmental related activities in various parts of the world (e.g. EIA programmes). FEPA as we pointed out earlier was set up by Decree number 38 of 1998 as amended by Decree Number 59 of 1992 and has overall responsibility for environmental protection in nigeria. It has both advisory and executive powers, which include establishing of state environmental protection agencies. Mechanisms for the application in Nigeria of all international conventions and protocols on the environment to which Nigeria is a signatory. It is also the Designated National Authority (DNA) for Nigeria to international operation instruments to which the country is committed. FEPA's three main operations organs or structures are: The Governing Council, the National Council for Environment and the Conference of General Managers/Directors of State Environmental Protection Agencies that it works with to ensure a clean and pollution free Nigeria as well as ensure other environmental-friendly policies, projects or activities (Adegoke, 1977).

The question we must ask at this point is to what extent has FEPA achieved its objectives or met its responsibilities. Your answer is as good as ours; not much. Good policies, poor implementation, are the bane of most (if not all) Nigerian agencies at the moment.

The Voluntary Environmental Groups (VEGs)

The VEGs refers to organizations or groups who engage in advocacy and other concrete activities aimed at directly influencing environmental friendly changes or government and other agencies' policies that will help to bring about those changes. They include many environmental-related NGOs (Non-Governmental Organizations), the Chamber Movement led by the Nigeria Chamber of Commerce, Industry, Mines and Agriculture (NACCIMA), private media organizations, some labour unions and others.

The Community-Based Environment Protection Programmes (CEPP)

CEPPs refer to different communities or community based groups that come together and work together, sometimes with the help of government and non-government organizations and sometimes without these, to embark on activities that will help to keep their immediate environments clean, safe and habitable (e.g. waters, environmental sanitation etc.). These are waiting to be mobilized, equipped, motivated and enlisted in the war against environmental degradation or pollution at the grassroot level. But they have not. This is, for sure, a major challenge for EPR managers, practitioners or consultants. They should be treated as veritable publics and partners in any serious EPR campaign in Nigeria and other developing countries.

This great need for us to partner with community groups and let them participate actively in any environmental protection campaign was endorsed by the World Bank's 1992 World report, which among other things emphasized "the importance of involving local people in setting up environmental priorities and in implementing sustainable environmental improvement ... opens door to a more participatory approach to the design and implementation of projects ... one that draws upon the strengthens of local governments (and local communities) to mobilize political support for on-going action" the following were then reported as example of successful community-based environmental projects in various parts of the world: *(Urban Age Journal, 1992)*:

1. The organic low cost sanitation system in Karachi, Pakistan.

2. The waste recycling enterprise of the Zabbaleen Community in Cairo, Egypt.

3. The Community latrines in low-income areas of Accra, Ghana.

4. The Dar-es-Salam 10 cell unit system on environmental problems in Tanzania.

Note that Nigeria is conspicuously absent in the above list of success stories in CEPPs, even though its compilers said it was not exhaustive, they must have picked and reported really outstanding and successful ones. Half successful, fairly successful and abjectly unsuccessful CEPPs will never be noticed and will definitely not be reported in world class environmental and other journals like the *Urban Age Journal.*

Communication, Media and Environmental Protection

One other vital fact the EPR managers must always remember is the great power of communication and media forms in influencing any form of development including the achievement of the objective of sustainable environment at any level: global, national or communal. They do not only serve as veritable means of reach, visibility, information, education and persuasion, but provide surveillance of the environment, correlation of the components of the society in responding to the environment, transmit socio-cultural heritage and help to force important or salient issues, policies, problems, onto the public agenda (Nwosu, 1990). Furthermore, we have demonstrated in other research-based published works that "the media can work along with a nexus of other interpersonal and group communication modes (traditional and modern) in multi-media scenario to influence people's development" oriented attitudes, actions and behaviours, and thus help to bring about development" (Nwosu, 1994).

There are definitely many more reasons that can and have been given on the very crucial role of communication and the media in environmental protection or management efforts. Several of them can indeed be found in some other chapters of this book. But, the point being made here and which must be noted is that no EPR manager can achieve any reasonable success without effective communication and media means, which are also well discussed in other chapters of this book. They must also know how to work well with journalists or media practitioners, because they are very important partners in the crusade for a much improve and sustainable environment. In fact, they know exactly what role to play in handling environmental issues and problems either alone or in partnership with other professionals such as EPR managers.

Hence one of them, Mohammed Abdel Aziz of Al Haram in Cairo, Egypt, has written what he considered to be the role of a news reporter or journalist in environmental issues and problems. According to him these include:

1.   To keep a continuous effort to inform the public and to make constant effort to increase this awareness about the environment.

2.   To rewrite or re-edit the technical and highly specialized scientific information (about the environment) in a simple style suitable for the public or audience.

3.   To produce detailed reports of environmental issues and problems through investigative reporting (Hester et al, 1987:95-99).

Fair enough. But to this list must be added other journalistic genres or media fares like editorials, radio commentaries, advertorial, features, articles, precision stories and supplementaries, talk shows or programmes on radio and television, vox pops and many other items on the environment. To what modern media can do, the EPR manager and media men should not forget to add the traditional communication media, inter-personal and group communication media and, of course, the new media provided by the Information and Communication Technology (ICT) such as the world wide web, the Internet etc., as and when these are available and accessible to the media men, EPR managers and the target public of any EPR campaign, and depending on the budget available to the EPR manager.

In addition the mass media practitioners in Nigeria and Africa must, like their counterparts in the developed countries, practice BEAT REPROTING which refers to the assignment of specialized and well trained journalists or reporters to specialized beats like the environment. This makes for great expertise on their parts, deeper knowledge of issues and problems in the environmental beat, proximity to the environmental news and events by being there all the time, greater contacts and closer relationship with the news sources on the environmental beat reporting (Nwosu, 1996).

To facilitate the beat reporting process, the schools of communication in Nigerian and African universities should offer specialized courses in environmental reporting, while the governments and other agencies, including the mass media themselves should hold periodic intensive and interactive workshops on environmental reporting or the environment and the mass media. Beat reporters on the environment should also grow fast into environmental activists by practicing advocacy journalism, as opposed to mere bald or naked objective reportage of environmental facts or the opinions of others on the environment, just the way they were received. This objective reportage approach may provide information and create awareness, but solid or adequate environmental reporting needs to go beyond this. And they can go beyond this by using advocacy journalism technique to crusade or advocate for best practices or change of behaviours on the environment. These can come in form of well researched and well-written pungent editorials, commentaries, articles, features and such media fares.

This is development communication (DEVCOM), development support communication (DSC) and behavioural change communication (BCC) in action, which are all under the much talked about theory and practice of the mass media or communication (Nwosu, 2003). To be more effective in BCC or DEVCOM or DSC as beat reporters on the environment, these media practitioners should be "EMBEDDED" in the environmental beat just like the American and British reporters were embedded with their soldiers in the Iraqi war fronts during Gulf War II or the second inversion or "Decapitation" attack of Iraq in April 2003 which saw the overthrow of Sadam Hussein and which helped their media to give blow by blow reports of the war and very in-depth analyses of the issues, people and events involved in the war, based on firsthand information/observations, deep insights and deep communication. A professional combination of the above recommended advocacy journalism, beat-reporting and the "embedded" strategies will help media in Nigeria and Africa to ensure that their target audiences,(Nigerians or Africans) do not engage in what Daniel Boorstin (1962) described as "a change of face rather than a change of heart", as far as positive behavioural changes on the environment which is our concern in this book, is concerned.

One other very important reason-why the mass media should adopt the above recommended approaches to handling the reportage of environmental issues and problems in Nigeria, especially the advocacy, beat reporting and the "embedded" approaches, is that it will help them to not only provide regular adequate and up-to-date information and opinions on the environment to the citizens, but to influence their environmental actions and behaviours, also combat the anti-environmental views of most industrial or corporate giants, captains, groups or cartels that have consistently opposed the flow of information to the citizens on environmental issues and problems. In the views of these industry or corporate leaders, provision of public access to environmental information should not be allowed because:

a.   The publics are not competent to interpret environmental data; data may be misinterpreted and create unwanted alarm. *(Journal of Environmental Planning and Management, 1994).*

Our response: This is a baseless and wrong assumption because not all citizens are environmental illiterates these days and it is indeed the job of EPR and media managers to manage these environmental information to avoid any "unwanted alarm" and other such problems.

b.   Disclosure may confer a commercial advantage on competitors'.

Our response: This is as display of the typically myopic arrogance and insensitivity of business moguls which make them to sacrifice humanity or corporate social responsibility on the altar of profit maximization, competitive advantage or such pecuniary-influenced bottomlines. Quite untenable.

c.   The administrative Cost of Providing access would be out of proportion to the benefit, which would accrue.

Our response: Again, our response will be same as in (b) above. Hard-nosed and unrepentant businessmen and women would again sacrifice life on the alter of profit, pecuniary benefits and cost implications. This is business without a human face.

d.   The history of attempts to promote public participation in planning suggests that the level of public interest in environmental information is low.

Our response: Again, this is untenable because it is not 100% or a watertight correct statement or assumption. Moreover, these businesses or industries have a CSR duty to work with EPR mangers and media managers to raise and keep raising this supposedly low level of public interest in environmental information.

e.   The release of information to the public may harm the environment by, for example, attracting people to nature conservation habitats.

Our response: A very weak argument indeed that is supported with a weak example. Is it an impossible task to check people's negative attraction to the conservation habitats, and should not positive and enlightenment-oriented attraction of people to nature conservation habitats not be encouraged by all means and at all cost, if we sincerely believe in environmental protection or conservation?

It is therefore in line with our above-stated responses that we shall in this book strongly agree with or endorse the following reasons given by the *Journal of Environmental Planning and Management* (1994) as the case in favour of public access to environmental information;

a.   It will reassure the public and promote their participation in the actions being taken by the government and the industry (the public assurance role of EPR).

b.   The knowledge that activities will come under public scrutiny should act as a vital discipline for environmental protection agencies (the agency accountability role).

c.   Increased scrutiny will encourage industry to take environmental protection seriously (the industrial or corporate responsibility role).

d.   It will enable those outside the statutory enforcement system to play a part in safeguarding the environment (the public participation role).

Some Research Findings on the Media's Role in EPR

In working on this book, most of the systematic studies or researches we found on environmental issues and problems in Nigeria were general ones done mainly by experts in environmental studies. They include the ones by Aina (1990); Ukpong (1994); Umeh and Uchegbu (1997); Uchegbu (1998); Mba et al (ed, 2001) and others. We found very little research published works done directly on environmental public relations, except very few review professional journal articles like the one by Onumuonu (1986) which used no definite and rigorous or systematic research methodology. What we found is a fairly good number of published management of environmental issues and problems. We shall summarily review some of them in this chapter because of their direct relevance to EPR. As we know, the mass media are important tools, which EPR managers must work with at various stages of the communication components of their work.

The first attempt to study or analyse the role of the mass media in environmental management in Nigeria was in 1983 at a national workshop organized in Lagos by the Federal Ministry of Housing and Environment. Unfortunately none of the papers presented at the workshop applied any of the quantitative social science research, behavioural management or research methods like survey, content analysis, experimentation or systematic participant or non-participant observations. Many of the papers, however, presented useful insights on the subject matter or theme of the workshop, which was "Environmental Awareness for Media Practitioners" One of such insightful papers was the one by Andrew moemeka (1993) which among other things, pointed out that based on facts available in the literature in other parts of the world and in other areas of human existence in Nigeria, the "mass media have the capacity to inform, to educate, to arouse and to create awareness, and when they are imaginatively used, they do serve as effective channels of communication", including of course communication on or about the environment.

The workshop also revealed that the number of stories (news, articles, editorials etc) published about the environment in Nigerian mass media was quite low. The workshop could not confirm however, whether the low coverage of environmental issues by the Nigerian media was due to low awareness level of these problems and issues among the media practitioners, insufficient knowledge on their part (on these issues), or insensitivity to the environmental issues or problems. Luckily, one of the resource persons at the workshop designed and administered a questionnaire to the media practitioners who participated in the workshop and later published his findings (Nwuneli, 1985). Among other findings, his study revealed that none of the media practitioners studied had ever attended any environmental awareness or training workshop in or outside the country, and none of them had ever been assigned to cover any news event or source on the environment. The surveyed media practitioners also reported that their media houses had no policy on the coverage of the environment, and their editors did not have enough knowledge of and interest in the environment to consider its reportage or coverage as important.

B.A. Choker (1985) also carried out a survey of articles, editorials and other materials published on the editorial pages of three Nigerian newspapers *(The Daily Times, New Nigerian and Concord)* to find out how their publications on environmental issues and problems compared with other issues and problems. His study showed that over 18% of the published materials in *Daily Times* was devoted to economic issues; while the *Concord* devoted up to 24% to the same economic issues. In contrast, environmental commentary in the *New Nigerian,* was about 8%, 5% in the *Concord* and 7% in the *Daily Times.* What gaping contrast; what a wide gap! ... a clear indication that Nigerian media and media practitioners were in the mid and late 1980s paying little or no attention to environmental issues and problems.

What about the 1990s? Did the above unfortunate situation really improve? Not really. Studies done within that period confirm this fact. For example, Ikechukwu Nwosu (1997) studied two government newspapers *(New Nigerian* and *Daily Times)* and two privately-owned Nigerian newspapers (the *Champion* and the *Guardian*), quantitatively using the systematic content analysis research method. This study which was published in a professional public relations journal (see references) because of its importance to EPR and which is reproduced in chapter Eleven of this book as a case study, revealed among other things, that even up to the late 1990s the mass media in Nigeria did not only quantitatively undercover environmental issues and problems, but gave low quality coverage to the little they published. In 1999, Nathaniel Anokwute (1999) studied the same newspapers analyzed by Nwosu above *(Champion, Guardian, New Nigerian* and *Daily Times)* and concluded, after systematic content analysis of the papers, that the Nigeria newspapers did not give adequate coverage and analysis of environmental pollution issues and problems.

What about the performance of the Nigerian mass media in covering environmental issues and problems in the 2000s or the present millennium? The findings of researchers while reporting some relative improvements still end their research reports on cautious notes and expression of the need for further improvement by the media in Nigeria in this critical area of the media business, national and global survival and development. One of such studies which was done under the supervision of one of the co-authors of the present book (Ikechukwu Nwosu) was carried out by Emmanuel Ebinne (2002).

The study was a scientific content analytical case study of 110 selected ecological issues in four Nigerian Newspapers and for a 12 month period. It reported that "Out of the 110 stories. *This Day* had 41 stories or 45.1%. *Daily Times* followed with 35 times or 38.5%. The *New Nigerian* published 25 times or 26.4%, while *the Champion* published only 10 or 11%, of the published items.

This study also noted that there has been a relative "improvement and balance in key ecological sanitation and degradation reflecting 30 stories, and global warning which got zero coverage in previous studies recording at least 5 stories in this study". The researcher observed that the above pattern of his research findings show "that the media practitioners are beginning to come alive to their responsibilities. This could be attributed to the growing awareness on ecological issues created globally and locally by individuals and organizations". He was however, quick to add that there is therefore, the room for the media "to perform better".

We cannot agree more but will like to add that if all the parties involved in environmental management in Nigeria do not maintain and indeed heighten the tempo of their environmental activities, the above reported trend of hope or improved performance of the media in this area in the current millennium, will surely slacken. This is part of our reasons for embarking on this book project which we hope will help to catalyse this process of continuous awareness and knowledge of environmental issues and problems, which will in turn lead to continued good performance by the mass media, and also by all other stakeholders that include EPR managers, governments, NGOs and the general public.

Conclusion

Knowledge is power and application of knowledge even more powerful. Thus environmental knowledge is environmental power and so is application of this knowledge very potent. EPR managers, media practitioners, the general public and all other stakeholders of the Nigerian environment will do well to keep expanding and updating their knowledge base on environmental issues and problems in the country and the world. This will help all and sundry to contribute their desired quota to the establishment and sustenance of a healthy environment in the country.

Everyone, it must be noted, has a role to play. This is because the environmental problems call for multidisciplinary, interdisciplinary and multifaceted approaches to problem solving, before results can be achieved. Adegoke Adegoroye (1997:13) would agree with this observation, hence he has publicly pointed out that transforming Nigeria to a nation of environmentally conscious citizenry where industry, developers and business would generate in the best ethics of environmental responsibilities, requires not just governmental and professional bodies alone, but the coordination and commitment of environmental non-governmental organizations.

(ENGOs), Community Based Organizations (CBOs) and Social Association (SAs). Adegoroye who was speaking as the Director-General and Executive of the Federal Environmental Protection Agency (FEPA) stated that by 1997,

> Our register of ENGOs in FEPA shows that we have over 60 of them in Nigeria. Notable among these are: The Nigeria Conservation Foundation (NCF), the Nigerian Environmental Study/Action Team (NEST), the Friends of

the Environment, the Abuja Horticultural Society and the Nigerian Environmental Reporters Association.

Very importantly, the professionals should endeavour to not only keep abreast with local and international environmental issues and problems, but must infuse them into their professional actions, decisions, and policies. This is because it has been observed that "for many of our professions and disciplines non-infusion of environmental judgements make them out of tune with the current dictates of sustainable development" (Adegoroye, 1997:10).

In addition to the ones discussed in this and other chapters of this book, the local or specific Nigerian environmental problems and issues that must be known and monitored frequently by Nigerian professionals, especially EPR managers are stack fumes from industries, municipal wastes, nuisance, non-biodegradable petrochemical products, litters in the cities, crank case oil disposal from mechanic workshops, unsettled dump sites improperly reclaimed for development, invasive species like water hyacinths, refuse "jungles" that litter Nigerian cities like Lagos or even Abuja due to poor urban planning, health related environmental problems in the urban and rural areas (e.g. typhoid, cholera, guinea worm attacks etc), establishment of industries or manufacturing companies in residential flats or areas in urban town like Lagos, Nnewi, Aba etc, and digging wells near toilet sewages, factory effluents, air pollution, chronic stress or the chronic fatigue syndrome (CFS).

The global issues they should also keep monitoring include continual global climatic changes, ozone layer depletion, global warming, green-house effects and trade-related international environmental problems that include trans-border movement of toxic chemical waste, and contraband chemicals and pesticides with daring strategies approaching what can be described as *toxic terrorism* (Adegoroye, 1997:3). With respect to global warming, we must understand that, its theory is usually backed by two assumptions: that the melting of glaciers and ice caps will increase the amount of water, and that sea water expand when warmed.

Environmentalists in the developed countries were disturbed that the melting of ice in certain regions could have devastating effects on animals that depend on ice packs such as ringed seals and polar bears. They were also worried that warming could result in loss of plant diversity even in the relatively safe tundra ecosystem. Global warming is caused mainly as a result of the build-up of greenhouse gases such as carbon dioxide (CO_2) (Omole, 1997:13).

Greenhouse effects are caused by the entrapment of some earth radiated heat in the atmosphere. This entrapment is usually enhanced by gases in the atmosphere, which act like glass in a green house. The green house process is created thus: sunlight penetrating the atmosphere warms the earth's surface; the earth's surface radiates heat to the atmosphere and some escape into space; greenhouse gases and water vapour absorb some infrared wavelengths (radiated heat) and reradiate these heat back to the earth as a consequence of the greenhouse gases build up in the atmosphere, more heat is trapped near the earth surface. Ocean surface temperatures rise; more water vapour enters the atmosphere and the earth's surface temperature increase (Miller, 1995:213).

Chapter Six

Degradation And Environmental Public Relations: The Past, Present And Future

Introduction

The earth is at risk as never before. We have contaminated lands and rivers with poison, fouled shorelines and oceans with massive spills of oil and altered the chemistry of the air on which life depends (Porrit, 1991:92). People today tend to measure progress almost exclusively in materialistic terms. But it is good to note that there are now evident unanswerable cases that such "vision" if pursued to its logical conclusion, must inevitably destroy that which it set out to achieve. The aspiration to go on getting richer, come what may in ecological terms, must inevitably destroy the natural resources and life support systems on which we depend.

Stressing this point of view further, there has been a scientific consensus about the environmental problems plaguing our earth. On November, 18, 1992, some 1,680 of the world's senior scientists from 70 countries, including 102 of the 196 living scientists who are noble laurettes, signed and sent an urgent warning to government leaders of all nations. The warning runs thus:

The environment is suffering critical stress ... Our massive tampering with the world's interdependent web of life coupled with the environmental damage inflicted by deforestation, species loss and climate change-could trigger widespread adverse effects, including unpredictable collapses of critical biological systems whose interactions and dynamics we only imperfectly understand. Uncertainty over the extent of these effects cannot excuse complacency or delay in facing the threats ... No more than one or a few decades remain before the chance to avert the threats we now confront will be lost and the prospects for humanity immeasurably diminished ... Whether industrialized or not, we all have but one life-boat (The earth). No nation can escape injury when global biological systems are damaged ... We must recognize the earth's limited capacity to provide for us, (Miller, 1998:25).

Also in the same 1991, the prestigious US national Academy of Sciences and the Royal Society of London issued a joint report, their first ever, which reads thus:

If current predictions of population growth prove accurate and patterns of human activity on the planet remain unchanged, science and technology may not be able to prevent either irreversible degradation of the environment or continued poverty for much of the world. Thus, all these point to the urgent need for a re-orientation of the people towards a sustainable development (Miller, 1998).

It has been observed that pollutants can enter the environment naturally (for example, from volcanic eruptions) through human (anthropogenic) activities (for examples, mining, burning of coal, improper disposal of refuses and sewages, contamination of streams, and drinking water sources, unrestricted disposal of industrial effluents and wastes). Equally, most pollutions from human activities occur in or near urban and industrial areas where pollutants are concentrated. Some pollutants contaminate the areas where they are produced; others are carried by wind or flowing water to other areas. Pollution does not respect local, state or national boundaries.

Furthermore, some pollutants come from single, identifiable sources, such as the smokestack of a power plant or the exhaust pipe of an automobile. These are called *point sources.* Other pollutants come from dispersed or non-point sources. These are often difficult to identify. Examples are run-off of fertilizers and persticides from farmlands, golf courses and suburban lawns and gardens into streams and lakes. It is much easier and cheaper to identify and control pollution from point sources than from widely dispersed non-point sources. Morestill, three factors determine how severe the harmful effects of a pollutant will be.

1.   *The chemical nature:* This implies how active and harmful it is to living organisms.

2.   *Its concentration:* This means the amount per unit of volume or weight of air, water, soil, or body weight. (For example a concentration of one part per million (1ppm) corresponds to one part pollutant per one million parts of gas, liquid or solid mixture in which the pollutant is found). We note here quickly that one way to lower the concentration of a pollutant is to dilute it in a large volume of air or water. But currently, with the overwhelming of the air and waterways with pollutants, dilution is now but only a partial solution.

3.   *The pollutant's Persistence:* This entails how long it stays in the air, water, soil or body. We have Degradable or non-persistent pollutants. Degradable pollutants can be broken down completely or reduced to acceptable levels by natural physical, chemical and biological processes. Slowly degradable pollutants include insecticides, DDT and most plastics while Non-degradable pollutants cannot be broken down by natural processes. Examples are toxic elements as lead and mercury and some other effluents from industries. The best way to control this kind of pollutants is not to release them into the environment at all or to recycle or reuse them. Removing them from contaminated air, water or soil is an expensive and sometimes impossible process.

The major environmental problems include:

a.   *Air Pollution:* This come in form of or results to Global climate change.
     Stratospheric ozone depletion
     Urban air pollution and acid deposition

b.   *Water Pollution:* Forms of this include
     Oil spills
     Toxic chemical
     Pesticides
     Infection's agents
     Nutrient overload
     Sediments

c.   *Waste Production:* This entails
     Solid waste and refuses and
     Hazardous waste.

d. *Biodiversity Depletion and land Pollution:*
    Habitat destruction
    Forest depletion
    Habitat degradation
    Extinction

e. *Food Supply Problems*
    Farmland loss and degradation
    Soil erosion
    Soil salination and soil water logging
    Water shortages
    Potable water contamination

Each of these forms or types of pollution has chains of agonizing environmental consequences that are far reaching even unto generations still to come. For example, Depletion of the ozone layer which is an outcome of air pollution, damages the DNA in humans and causes acute crythema (sunburn) and keratitis (snwo-blindness); chronic cataracts and skin cancer. It equally decreases immune surveillance and thus increases susceptibility to cutaneous infection and air cinogenesis (Jones, 1989:208). As regards effect of Depletion on plants and marine organism; here is an abstract by Rovert C. Worrest and Dester D. Grant of US Environmental Protection Agency:

> Experimental evidence suggests that increased exposure to ultraviolet – B Radiation (an outcome of Ozone Depletion) at the Earth's surface, would have negative effects on both terrestrial and aquatic biota ... Crop yields are potentially vulnerable ... forest productivity is also affected ... It will modify the distribution and abundance of plants, and thereby change ecosystem structure ... it equally causes damage to fish larvae and juveniles and other small animals

and plants essential to the marine food web. Effects include decreases in reproductive capacity, growth, survival and other functions (Worrest and Grant, 1989:197).

On another note, global warming which was caused by pollution and by greenhouse gases would lead to rise in sea level and culminate in loss of wetlands, pollution of estuary waters, decrease in soil moisture, and increased tropical storms (oppenheimer, 1989:100). Morestill, *the Guardian* of December 16, 1997 reports that global warming would enhance the El-Nino phenomenon and its effect (Etuk, 1997). Furthermore, Houghton maintains that with the warming, the activities of many insect disease-carriers, which thrive better in warmer and wetter conditions, would double. For example epidemics of such diseases as viral encephalitis carried by mosquitoes would increases (Houghton, 1994: 84). Equally, some diseases currently confined to tropical regions with warmer conditions could spread into mid-latitudes, Houghton observed.

As it is, there are only two basic ways or approaches to dealing with pollution. Prevent it from reaching the environment or clean it up if it does. Pollution prevention or input pollution control slows or eliminates the processes. Pollution can be prevented by switching to less harmful chemicals or processes, either through reuse, recycle or nominal reduction. On the other hand, pollution clean-up or out-put pollution control involves cleaning up of pollutants after they have been produced. But environmental scientists have always urged that we emphasize prevention because it works better and it is cheaper. As the age-long saying will maintain, an ounce of prevention is better than a pound of cure.

Pollution clean-up has three major disadvantages: firstly, it is often only a temporary bandage as long as population and consumption levels continue to grow without corresponding improvement in pollution control technology. Secondly, it usually removes a pollutant from one part of the environment only to cause pollution in another part. For example, we can collect gabbage, but at the end, it is either burne (perhaps causing air pollution and leaving a toxic ash that must be put somewhere) or dumped into streams, lakes and oceans (perhaps causing water pollution). Thirdly, once pollutants have entered and become dispersed in the air and water (and in some cases, the soil) at harmful levels, it usually costs too much to reduce them to acceptable concentrations (Miller, 1998:18).

But the big question is: How would the world's populace come to embrace this pollution prevention or input control? They need to be enlightened and educated on the inescapable dangers of the scourge of environmental degradation and pollution and on the urgent need for a change of attitude and behavior towards our dear earth. This is where Public Relations is strictly indispensable. Hence the appraisal of the PR RICEE model as a strategy towards winning this enlightenment campaign.

Environmental Pollution

So there are four major types of environmental pollution in the world today. These are land, water, air and noise pollutions.

Land Pollution

Land pollution which affects over 40 million acres of land in Nigeria occurs when some pollutants are introduced into a piece of land or through alteration in such a way as to render it unsuitable for its best zoned uses. It is usually caused by such things as refuse dumps or scattered waste materials, rubbishes from demolition, unstable stripped solid, exposed erodible soil, rock from mining operation, junked materials, waste or spilled oils, soil cutting caused by quarrying operation and others. And the effects of land pollution include breeding of disease carriers (e.g. rats, flies and mosquitoes) increasing run-off erosion in flooding, killing of valuable or rare vegetation and wild life, destruction of aesthetics, production of bad odour and litters, ground and surface water contamination etc (Umeh and Uchegbu, 1997).

But according to Ukpong (1994) the major cause of pollution and degradation in Nigeria are improper resources management, destructive logging of our forests, overgrazing and over cropping of arable lands, flooding and wind erosion, over population, poor property management, poor sewage and waste management, poor law enforcement, indiscipline, inappropriate use of technology in farming and manufacturing, mining in Jos and Enugu areas, land degradation with pesticides and fertilizers, some known natural land slopes, and destruction of oil wells and marshes. To this must be added the serious effects of desert encroachment in far Northern Nigeria and the Atlantic Ocean encroachment in Lagos and other littoral areas of Nigeria. other serious areas of land pollution in Nigeria include the lead and other mining areas of Ebonyi State, (e.g. Abakaliki), the Okpella mines in Edo State, gypsum mining areas in Sokoto, the gold ore mining areas of Kebbi State and the limestone quarrying areas of Ewekoro in Ondo State.

These are somewhat similar to the causes of land pollution in the developed countries like the U.S.A. and Europe as identified thusly by Tyler G. Miller (1991): Periodic flooding, water logging of soils, nutrient depletion of soil, and washing away of soil and crops, cultivation of land, deforestation, overgrazing and mining, urbanization, agriculture,logging and construction, all of which can erode the soil and so degrade or pollute it.

Water Pollution

Water pollution occurs when a concentration of certain pollutants are introduced into any water or water source for reasonably long period for it to have some negative effects on the physical, biological and chemical qualities of water. In Nigeria, water pollution is usually caused by water pollutants that include coloured matters, heated liquids, organic matter, mineral salt, detergents, toxic chemicals, industrial wastes, domestic waters, and oil spillages. Water pollution has negative effects that include causing or increasing of corrosion of surfaces which the water comes in contact with, encouraging the growth of undesirable biological life in excessive quantities; interfering with the quality of water for drinking, bathing, boating recreation etc. and rendering the water unfit for industrial irrigation, and domestic purposes (Umeh and Uchegbu, 1997:53-54).

Water pollutants from industrial plants or companies (e.g. manufacturing firms) include cynide, lead, sulphates, nitrates, arsenic and many other such dangerous and poisonous elements.The dangers that they pose to man, animals, and living organisms are quite enormous. According to P.A. Vesoland and J.J. Pierce (1982) the main causes of water pollution in the developed areas of the world are as follows: organic water from industrial plants; organic wastes from industrial plants; unknown chemicals; heat from industrial discharges, municipal wastes, agricultural wastes; sediments from land erosion; acid rain, oil spills and contributions from routine human operations or activities. These are no doubt not really different from the causes of water pollution in Nigeria and other developing countries of the world. But in Nigeria and other oil producing countries, oil spillage, is definitely one of the most outstanding causes of water (and even land) pollution especially in the riverine and other oil production areas like the Niger Delta areas of Nigeria which include Rivers state, Bayelsa State, Delta state, Akwa Ibom State, Abia and some parts of Imo State, Edo State, Ondo state. For example, a single reported case that occurred in 1980, the Funiwa Number Five oil well blow-out resulted in the loss of 400,000 barrels of crude oil and the death of large numbers of fishes, crabs and plants (Nwankwo and Ifedi, 1980).

And recent statistics from the department of petroleum resources (DPR) of Nigeria, shows that the above reported 1980 oil spill is in no way an isolated case and that indeed the trend from 1980 to the present day shows continuous upward movement in the cases of oil spills in Nigeria on the oil spillage chart or frequency polygon. According to the DPR, (in *This Day*, 2001), the volume of oil spillage has been on the increase since the early 1990s; from 14,940 barrels in 1990 to 98,345 in 1998, then, from 84,072.13

barrels in 2000 to 120.970, 16 barrels in 2001. The DPR went further to point out that 80% of the total quantity of oil spilled was due to equipment failure.

Even though the DPR has stated that most oil spills in Nigeria were due to equipment failure and metal fatigue, largely because integrity tests on oil pipelines were not carried out as and when due, a reasonable number of oil spills were attributed to sabotages that come in such forms as willful destruction of pipelines and oil producing equipment, as well as community crises which often lead to vandalization of oil producing sites, pipelines and equipment. All these go to show that EPR managers in and outside the oil companies have a lot of work to do in this regard. For one, there is need to advise, remind, and ensure that the oil companies inspect pipelines regularly, police and monitor them always to avoid their vandalization (in collaborating with the communities and the governments), step up EPR community relations activities that will educate the angry and jobless youths in the oil producing communities, who carry out these vandalization most of the time, that their activities do more harm than good in their communities, and so they need to dialogue more frequently with these oil companies and the governments, instead of taking the laws in their hands. The EPR manager and his team must also use PR issues and crisis management approaches (Nwosu, 1996) to help in managing the usual crises, which these situations bring about. Perhaps, a look at the following four cases of recent oil spills reported in Table 6.1 below will help to drive home the points we have made above on the deleterious and devastating impact of oil spills on the environment:

Table 6.1:   **Some Recent Oil Spills in the Niger Delta**

(From NDDC's EP & C's 2001 third quarter report)

	SPILL INCIDENT	DATE	COMPANY INVOLVED	CAUSE OF SPILL	HABITAT IMPACTED
1.	Qua Iboe terminal Tank farm splliage	16/5/2001	Exxon Mobil	Pressure surge, valve Opened due	2½ km stretch of coastline impacted by crude oil
2.	Oil spillage and explosion at Umudike II in Ohaji Egbema Local Government Area	3/11/2001	SPDC	Equipment failure	Farmland, Homes, Vegetation
3.	Fire outbreak and pipeline vandalization at B-Dere and K-Dere Communities in Gokana Local Government Area	25/8/2001	SPDC	Sabotage several kilometers of the trans Niger pipeline were excavated and cut in sizeable lengths for onward transportation to buyers	Swampy Basins and Vegetation which consists secondary forest and farmland
4.	Quo Iboe terminal Spill in Mkpanak	8/8/2001	Exxon mobil	Sabotage suspected ½" plug was removed	Grassland and some buildings in the vicinity of the spill point

Source: Ugochukwu, Onyema (2002:4) and NDDC's EP & 2001 Third Quarter Report.

Air Pollution

Air pollution has to do with dangerous contamination of the atmosphere, which reduces the quality of the air we breathe and which sustains other lives on earth (Nwosu, 2003). And the atmosphere has been described as the thin envelope of life-sustaining gases surrounding the earth" (Miller, 1998:1860).

Air pollution can be caused by nature and can be man-made. Some of the natural causes include volcanic eruption, whirl winds or wind storms, earth-quakes and others, while the man-made causes include wrong solid waste disposals, gas flames, oil exploration, production and use, industrial pollution, and many other such factors which also have damaging effects on both aquatic and terrestrial lives and contribute to biodiversity loss (Uchegbu, 1998:14). They also have serious negative effects on human health and materials. Other air pollutants like cigarette smoke, radioactive random – 22 gas, asbestos, aerosol sprays, some room deodorants, gasoline etc, cause dizziness, headaches, coughing, sneezing, burning of eyes, flu-like symptoms that is known as the sick-building syndrome.

To change the adverse effects of these air pollutants on man, experts advise, the regular change of air fillers, cleaning our air conditions system, exchanging humidifier water trays frequently, not storing inside our room gasolines, solvents or other volatile and hazardous chemicals. In addition, it is advised that we do not use room deodorizers or air fresheners and aerosol spray products, control our smoking habits and making sure that our wood-burning, gas or kerosene cookers or stoves and fire places are properly installed, vented and maintained.

We cannot talk about air pollution without discussing the GREEN HOUSE and its effects. We have defined the green house as a place, which traps heat in the atmosphere and have explained how it is actualized or formed in the concluding section of chapter five of this book. We need to give more details about the green house because of its importance and effects on the environment. First, we must note that without the trapped heat in the green house, the earth would be too cold for human survival. Second, that various human survival activities have increasingly and negatively poured into the atmosphere gases that capture too much heat eg. carbondioxide from burning of wood, coal, oil and natural gas, produces most of the greenhouse gas while trace gases like methane nitrous oxide and chloroflorou carbons (IFCS) continue to rise (*Newswatch*, 1990:10-11).

Experts have warned that if – the build-up of these greenhouse gases in the atmosphere is not halted, it will result in what we know today as GLOBAL WARMING which will bring about adverse drastic climatic changes that include the following: shift in rainfall pattern changes which disrupts agriculture in many areas of the world; rise in sea levels, which results in the flooding of the coastal or littoral areas (e.g. the Lagos bar beach or Victoria Island, Lagos); shift in ocean currents which affects the climate and fishing activities negatively; reduction of favourable habitats which will result in the extinction of many plant and animal species (e.g. in Madagascar and many parts of Nigeria or Africa), rise in heat waves, droughts, hurricanes and other weather anomalies that would definitely harm man, animals, crops, forests, etc (Gues Speth, in Miller, 1998:18).

Luckily, there are things we can do to avert the above-listed negative effects of global warming and the DEPLETION OF THE OZONE LAYER. Question is, are we ready to embark on these interventions? And, how many people, especially in the developing countries like Nigeria, even know these interventions or even what global warming means or portends. These are no doubts enormous tasks for the EPR managers who play key roles, as we have discussed in chapter two of this book in these aspects of managing the environment.

What are some of these possible interventions by men and organizations? They include banning all production and use of chloroflora carbons and halos; greatly improving energy efficiency to reduce emissions of carbondioxide; and other pollutants; shifting to perpetual and renewable energy sources that do not emit carbondioxide. Slowing population growth because if we cut greenhouse gas emissions in half and population doubles, we are back where we started; planting trees, recycling carbondioxide released in industrial processes; getting air-polluting old cars off the road, stopping gas flaring and indiscriminate bush burning (Miller, 1998:221). These interventions are mandatory for man now because according to the United Nations inter-governmental panel on climatic change (IPCC), the last two decades of the 20th century were the hottest on record (UN, 1998) as a result of global warming and depletion of the ozone layer. And as we all can tell, it is not getting any cooler today.

Noise Pollution

One often neglected area of environmental pollution today is noise pollution. As a result, many people do not know that there is anything like that or the negative impacts or consequences of noise on man and his environment today. In many Nigerian cities today (e.g. Lagos, Onitsha and Aba), noise has become a very big problem that need to be managed. But as noted above, most residents of these urban cities are quite ignorant of noise pollution and its effects or impacts. There is therefore, a big job for EPR managers in Nigeria, in this important area of environmental management.

Those who work in or attend disco clubs (e.g. the youths) in Nigeria are quite oblivious of the negative incremental or long-term effects of the heavy sounds that blare from the loudspeakers as well as the human noise (e.g. shouting or singing along with the music outputs, shuffling of legs etc) have on their overall health and learning particularly. The neighbours that live in such noisy places, including noisy motor-parks, can however tell you the big discomfort they have to cope with every day and even night. The only problem is that most of them feel they cannot do something about it and so have resigned to their fates. They can do a lot. And it is the job of the EPR managers and other environmental experts to let them know the injury they are causing to their health by remaining complacent, as well as actions they can take to save themselves and others from this avoidable environmental problem.

Those living close to or working in industrial plants or industrial areas also suffer immensely from noise pollution. They also need the attention or intervention of EPR managers and environmental experts through the managers of these industrial areas and plants. They should educate them to the fact that what constitutes noise as a part of environmental pollution management is usually loud and unwanted sound and that sounds are usually measured in decibels; the higher the decibel of any sound, the more qualified it is to be regarded as noise that is harmful to man and his environments.

Sound experts have classified sound types of 0 to 60 decibels as audible, quiet or moderate and therefore harmless, while sound types from 70 decibels to 150 decibels are described as loud, very loud, uncomfortable or even painful. Hence they have described threshold hearing of 0 decibel as audible. Normal breathing which usually has 10 decibels; leaves rustling in breeze which have 20 decibels usually as very quiet; whispering which usually produces 30 decibels as also very quiet; library which usually has 40 decibels is described as quiet; restaurant which has 50 decibels of sound as quiet; and normal conversations which produce 60 decibels of sound usually as being moderate. On the high and therefore noisy spectrum vacuum cleaners which usually produce 70 decibels of sound are described as loud. Food blenders with 80 decibels are described as very loud also; moving train which produces 100 decibels of sound is described as UNCOMFORTABLE; machine gun firing at close range which produces 120 decibels of sound described as extremely loud; and jet plane engine at take-off which produces 150 decibels of sound is described as PAINFUL (Paul R.E. and H.E. Anne, 1997, cited Uchegbu, 1998).

With all the above examples from the sound and noise experts, only God knows the cumulative negative effects all the above, coupled with the extremely noise and many hoe-installed electricity generators which are now common in many urban towns and even some rural towns and neighbourhoods, which serve as alternative to public power supply from the Nigerian Electricity Power Authority (NEPA), have been having on the mental, physical and other health dimensions of Nigerians. The danger is definitely on the very high side, considering the fact that because of the very low and epileptic supply of electricity by NEPA, most Nigerians including industrial plants, all depend more on their generator sets than the NEPA public power supply for many hours of each day when the public power supply is so bad or so poor in Nigeria that many Nigerians would confirm and have indeed expressed the fact in many fora, that NEPA is a standby power supply source, while their self-installed electrical generators are the major source of power. May God save us from the noise pollution and the other environmental pollutions e.g. frequent release of carbondioxidse and other dangerous gases, which the above stretched situation has brought about.

Regulation and Control

Again, even though we have discussed this, in part, in chapter five of this book and in some other chapters (tangentially) we want to discuss further this aspect of environmental management, regulation and control, because of its importance as an intervention method. Generally, there are three broad methods of regulating and controlling environmental activities in Nigeria. These are statutory provisions for prohibiting and controlling pollution as well as environmental degradation, the National policy on the environment and the oil spillage monitoring programmes (Mba, 2002:13-14). We have discussed the National policy on the environment in various parts of this book and have excerpts from it in Appendix I of the book. So, we will not dwell on it here. With respect to the Nigeria's statutory provisions in addition to the ones we have provided in chapters 5, other chapters and the Appendices, we like to also draw the reader's attention to the following laws: The Mineral Acts of 1946; the Mineral Oil (safety) Regulation of

1967; the Oil in Navigable Waters Acts, Number 34 of 1968; Petroleum Regulations of 1967; Petroleum Decree, (Act) of 1969; Petroleum (Drilling and production), Regulations of 1969; Petroleum (Drilling and production) Amendment Regulation of 1973; and the Petroleum Refining regulations of 1974. Nigeria also has statutory regulations against erosion-inducing activities, bush burning and quarrying in environmentally unsuitable areas (Mba, 2002:13-14).

But the problem with Nigeria is not making laws and policies. The problem is with implementation; so, we must ask and answer the question: To what extent have the Federal Environmental Protection Agency (FEPA), the National Council on the Environment, the state governments and other appropriate government ministries and agencies at the federal, state and local government levels, been able to apply these environmental laws, regulations and policies:? Your answer is as good as ours.

With respect to the third type of interventions or controls, the oil spill monitoring programmes or policies, the story is the same – abysmal failure. What are some of those programmes or policies? There is the monitoring programme set up by the Inspectorate Division of the Federal Ministry of Petroleum and National Resources to oversee, monitor and control the activities of the oil companies. But to date, its impact is still to be felt in the country, as they have been quite unable to control the oil spillage, and other environmentally unfriendly activities of the oil companies. Again, the failure story is the same for the three other specific programmes set up by the government to monitor and check oil spillage and other environmentally unfriendly activities of the oil companies. These are: the oil spillage controls programme with contingency for minor to medium scale oil spills; the efficient discharge control programme which was designed to monitor and control the impact of discharges on the environment; and the Environmental baseline studies programme which was set up by the Petroleum Inspectorate Division mentioned earlier, for collecting, collaborating, and anlaysing data on impacts of oil operation on the Niger Delta environment on a regular basis. These programmes have not lived up to expectation.

So, where do we go from here (Quo Vadis)? What needs to be done? The answers to the above two questions are scattered throughout the book. But at this point, we can only say that the "sleeping dogs" should be woken from their slumber by the environmental NGO's environmental activities, the mass media, EPR managers and any individual or group interested in saving mother earth from the environmental catastrophe that will definitely take place if we remain dormant, complacent, uncaring or keep doing business as usual.

From the above analyses we can easily see that as the poor earthlings that we are, we have been very unfair to mother earth from the Adamic age, the stone age and the present day. Whether it is land, water, air or noise pollution, we are as guilty as the most guilty criminal in terms of consistently destroying or exploiting the environment, without adding any value or replenishing any part it, all in the name of survival, growth and development that are in no way sustainable.

In every part of the world or mother earth, this story of "woes and crime" against the environment is the same. For example, while "every European leaves a life time of wastes one thousand times his or her body weight, for every person in the third world, forty square hectares of rain forests are destroyed" (Harrison, 1993). Take the sad and unfortunate case of Madagascar. That country is said to illustrate better than any other place on earth the creative processes of the past and the destructive forces of the present. The wild life is a living record of commercial drift. It has elements of Africa and India. Its long separation created new species and preserved primitive characteristics that died out elsewhere. But ugly men and women from within and outside Madagascar have been and had continued to destroy and rape that country's rich natural endowments.

Of the original forest cover of 11.2 million hectares, only 7.6 hectares remained in 1950. Today this has been halved to 3.8 million hectares. Every year 111,000 hectares more are cleared. At this rate all of Madagascar's rain forests will varnish entirely within thirty-five year ... In 1973 Ronomafana forest (in Madagascar) was 60 kilometres wide. By 1987, it has been cut back to a strip only 7-15 kilometres across" (Harrison, 1993:74-77).

And this immoral destruction of Madagascar's rich forests, wild life, biodiversity, arable land etc. continues even more aggressively today as reported recently by a CNN programme on the wildlife (CNN, November 11, 2003).

And this unfortunate trend is not restricted to Madagascar. The United Nations Environmental Programme has estimated that 6-7 million hectares of cropland are being lost worldwide each year, through soil erosion and another 1.5 million of irrigated and wet land to salination or waterlog. ... In 1992, global assessment of soil degradation found that between 1995 and 1990, almost 2 million square kilometers ... 17 per cent of the world's vegetated area ... became degraded... of this, 12.2 million square kilometers suffered serious degradation ... The highest share of degraded land was in Europe, 23 per cent; Africa came next with 22 per cent; then Asia with 20 per cent; in South America, 14 per cent were affected (Harrison, 1993:117).

In 2004, these figures would be definitely much higher because man is yet to learn the lesson of environmental protection, preservation and replenishment for sustainable development. This is more so because to date, the menace of soil erosion, brought about sometimes by nature but exacerbated by man's insensitive empowerment or development activities, has continued to take its toll on soil fertility and have very serious negative impact on crop or food productivity. In Nigeria, this truism has been demonstrated scientifically by the International Institute of tropical

Agriculture (IITA) located in Ibadan. On a test plot at IITA, "maize yields plummeted from 2 tonnes to 0.7 tonnes when the top 10 centimetres of soil were removed. When 20 centimetres were removed, grains field dropped to only 0.2 tonnes" (rattan; 1985:244). The data speak for themselves. But are we listening to them? Will they help to change our attitudes and behaviours towards the environment now? No, because we have not seen any evidence or sign of change coming from men and women in Nigeria and other parts of the world in this regard. This is despite the fact that (Paul Harrison 1993:119) has stated categorically, based on his many years of research on the environment, "that soil erosion is the world's most widespread form of land degradation".

Perhaps when we add the data or facts on the very harmful effects or impact of DESERTIFICATION on the global environment, some people somewhere in the world may start listening to us, at least, even if they do not take the desired environmental decisions and actions immediately or in the near future. And in doing this, we shall adopt the extended or wide-scope definition of desertification offered by the United Nations as most appropriate definition of this phenomenon. At the 1977 United Nations Conference on Desertification, it was defined as "the diminution or destruction of the biological potentials of the land which could lead ultimately to desert-like conditions" (Nelson, 1988). It is based on this definition, as opposed to the narrow definition of desertification as referring only to the rather unusual and infrequent desert advancement or desert encroachment, that the UNEP estimated that:

> Some 61 per cent of the world' productive drylands were moderately or severally desertified in 1985 ... In the developing countries, the proportion was as high as 77 per cent – including 80 per cent of rangelands, 75 per cent of rainfed cropland and 35 per cent of irrigated land. about 10 per cent of the rural population of developing countries

lives in affected areas – as many as 270 million people in 1990" (Mabbutt, 1991:103-113; and Nelson 1988).

Again, we must point out that these are frightening figures, especially when we know that at the time of researching, writing and producing this book, 2003/2004, which is more than ten years since UNEP gave these estimates, these figures must have jumped up significantly. Again this is man's inhumanity to mother earth, which contributes incessantly to the worsening of the world's environmental problems. Even in the relatively simple area of WASTE DISPOSAL/MANAGEMENT OR ENVIRONMENTAL SANITATION, man has continued to pollute and destroy the environment, with his wastes from Capetown- South Africa to Cairo- Egypt or from Lagos (West Africa) to Kenya (East Africa). And the matter is made worst by the fact that the rich elites in most societies, who produce most of the waste, have done next to nothing to dispose these wastes and ensure a healthy environment for themselves and the poor inhabitants of the earth. Paul Harrison (1993:257-258) and the *Environment and Urbanization Journal* were lamenting this immoral and insensitive behavior among the rich elites when they stated that:

> The rich use more resources wherever they are ... The rich discard more wastes. In PORT HARCOURT, Nigeria, the most profligate district throws away five times more municipal wastes per person than the least. Rich families contribute far more to global warming than poor.

This is not to totally exonerate the poor from the human "crimes" against the environment. We have all sinned and fallen short of the glory of God as far as environmental devastation is concerned. It is not therefore a question of casting blames or aspersions. We all need to work together, the rich and the poor (individuals or nations), to save mother earth.

The Future of Environmental Public Relations (EPR)

In the historical background offered in chapter one, we discussed the evolution of public relations right form when it was simply publicity to its present stage of development. But when we consider the future of public relations, we can vividly see the trend of public relations claiming the centre-stage in addressing immediate and even remote environmental issues and problems. The erstwhile Prime Minister of Canada, Brain Mulrney had an inclination to sharing this view when he foresaw two major issues that would be addressed by public relations in the future as health care and the environment.

As it is, there is this growth or increasing concern for the environment within various companies and organizations. This however, might have their roots in the need for different organizations to address the agitations against environmental pollutions from the indigenes or inhabitants of the community in which they are operating. Thus, this concern for the environment among various organizations could be viewed as a new initiative in the organizations' community relations and social responsibility programmes. (Nwosu, 2003; Nwosu, 2004).

The truth of the matter is that most often, taking adequate or proper care of the environment seems to reduce the optimal profit making of the organization, in that some part of the profit has to be some activities of the organizations, like indiscriminate disposal of industrial effluents are replaced by more efficient and environmental friendly means of effluent disposal, which we know, need some sizeable financial inputs. Hence, left on their own, most companies would have loved to continue maximizing their profits at the expense of the environment. But thanks to the agitations of the enlightened member of the communities and interventions by the governments, UN agencies and environmental agencies, which ultimately led to the global clarion call for sustainable development and environment. Examples of such agitations by different communities abound in Nigeria today, most especially in the areas of the South-south and South-east part of the country which are oil producing areas. Equally, another example where such conflicts of interest as regards the environment exist between an organization or even an industry and the publics, is seen in the dilemma of protecting endangered species (such as the spotted owl in Oregon) on the one hand and preservation of jobs and economic health (such as in the timber industry) on the other hand. Feelings run high on both sides, and the timber industry must combat a damaged public image, similar to the one that mining and oil and gas exploration companies had, before them, and so started repairing the environmental damage done during the course of their activities.

However, when we view the whole scenario from the long-term perspective, we are of the opinion that maintaining sustainable environment does more good than harm, not only to the community at large but even to the organizations themselves. We observe that if, for example, the company is disposing effluents in its vicinity, a point of saturation would be reached one day when the environment may not be able to take in any more wastes. This could lead to different forms of hazards ranging from epidemics to landslides. And when this happens, no one will be left out, the inhabitants as well as the company would have to suffer together. In fact, if care is not taken the company may be forced to fold up.

Secondly, maintaining a sustainable environment goes to project an organization as creditable and responsible. Thus, it helps to reposition the company's image and reputation positively in the mind of its publics. This in turn will help the organization to elicit goodwill, which it seriously needs for its surivial and growth. This is environmental public relations in practice. And we like to warn at this point that any company or country that pays a deaf ear to EPR now or in the future is digging its grave and so does not have a future. Its stakeholders, especially the external ones like the mass media, environmental activists, UN agencies on the environmental as well and the National one like Nigeria's Federal Environmental Protection Agency (FEPA), will indeed help it to die a much quicker death, if it does not develop and implement sustainable environmental-friendly policies. These are the present and future realities. To be forewarned is to be forearmed.

PART FOUR

****

ENVIRONMENTAL PUBLIC RELATIONS (EPR) IMPLEMENTATION STRATEGIES, MODELS

****

Chapter Seven

Environmental Public relations Practice and Management: Models, Strategies and Techniques

Introduction

World leaders were loudly warned as early as the 1990s of the impending environmental catastrophe. The warning came from the United States of America-based World Watch Institute. According to the institute's now widely known research report, among other things, "between 1972 and 1992, the world had lost 500 million acres of land, 200 million hectres of trees; that tens of thousands of plant and animals species had disappeared by 1972; same period; that lakes, rivers and even whole seas have been turned into sewages and industrial swamps; that air pollution has grown significantly, especially in most cities; and that unless these trends were halted within 20 years, the world's *biodiversity* would be drastically and dangerously damaged" (Nwosu, 2003).

The world leaders did not "concretely" respond to these warnings. What would have been regarded as a positive response by them was the very widely reported and now popular June 3, 1992 ten-days Earth Summit in Rio de Janeiro (Brazil) which was attended by 100 world leaders and 30,000 participants. This mere "talk-shop" weak response expectedly resulted in empty promises of concrete actions, winding motions and a Global Convention on ecological biodiversity, global warming and so on, that up to date have been signed by only 161 out of the 178 countries that participated in the summit. Even world super power and very wealthy U.S.A. is yet to sign and so ratify the convention, which has remained a mere paper tiger. So, world leaders have not really given the desired concrete or action-based responses to the warnings about the fast-approaching environmental disaster that may spell down to Earth and its inhabitants. This, no doubt, explains the key message of the 2002, June 5th World Environment Day theme which was:

> The Earth can be given a chance to survive only if and when the leaders (political, organization, industrial, community, professional, religious and other leaders) give their followers the chance to work side by side with them, or even in front of them (because he who wears the shoe knows best where it pinches), in the war against environmental degradation or destruction, through well-thought-out, well planned and well-executed *practical* environmental resuscitation, rejuvenation, protection and sanitation projects or programmes.

At the moment there seems to be a big gulf or total disconnection between the leaders and the followers on the search for methodologies or approaches and solutions to environmental problems and issues (Nwosu, 2003).

What seems to be urgently needed therefore, is to give the followers a true sense of belongingness, ownership, *partnership,* involvement or participation in our environmental management and protection projects. In fact, it will not be insensible to allow the followers to lead or show leadership in some stages of some environmental projects to give them a feeling of *mutual acceptance* between them and the leaders. After all, the projects have supposedly been conceived for them and in their interst and not of the leaders and experts who are usually in the minority. All these are in line with a popular slogan or maxim in development communication campaign management and execution which states that when followers *know* and *led* (or have some specialized knowledge and leadership skills) the leaders may not have a choice but to follow, to the mutual benefit of both and to the benefit of the entire society (Nwosu, 2003).

And this is where the relatively new but fast-growing area of management studies and practices, *Public Relations* (PR) must be brought in and applied maximally in any search for practical, realistic and sustainable solutions to environmental problems in the contemporary global society in general, and in developing societies like Nigeria in particular. The job of PR here will not only be to let the people, masses or citizens know enough about the environment, add the new knowledge to what they already know and be ready to show partnership oriented leadership; but to also build *"bridges"* that will make the followers to relate better with communities and organizations, and so work better with them as *one team* to find solutions to the problems of the environment (Nwosu, 2003).

Public Relations (PR) as a Vital Missing Link

We therefore, must all see public relations (PR) management strategies as the crucial missing links that must be added in the on-going attempts to reorientate all stakeholders in the search for solutions to our many environmental problems in the world, Africa, Nigeria and the Niger-Delta areas. And to do this, we have to first understand what public relations as a management function is all about, its functions, the three popular approaches to or methods (or types) of managing the public relations and PR-related strategies we can apply in any attempt to ensure the emergence of holistic, participatory and partnership-based strategies, that will help us to properly manage environmental issues and problems at any level of our global, national and regional existence. These are the major tasks of this chapters.

However, because there are now many books, journal, articles/reports on what public relations is or does, what it is not, its roles/functions, types/approaches (Nwosu, 1996; Black, 1990; Jefkins, 1989 etc), we shall only briefly discuss the above-stated dimensions of PR. We shall then concentrate in this chapter on what we consider to be the key PR and PR-related strategies that will be helpful to any organization, group, nation or the world in carrying our campaigns and actions aimed at protecting and preserving the environment.

Explaining Public Relations and the Three PR Types/Approaches

Stripped of all technicalities, public relations can be described as that function of modern management that uses series of well-researched, planned, systematic and sustained actions and communications to build and sustain mutual recognition and lasting *relationships* or partnerships between any organized group or corporate entity (e.g. an organizations, a country, a government etc) and its internal as well as external stakeholders or publics. Modern public relations (PR) emphasizes solid *image* and *reputation* building and sustenance based on solid performance or actions that are well communicated.

As a relatively new management function PRE is often confused with older professional practices like journalism, mass communication, marketing, propaganda, advertising, industrial relations, human relations and others that have some relationships with it. It is none of these. It is a distinct or autonomous management function and a full-fledged profession that at the same time draws from and contributes to the above-listed areas of management or professional practices (Nwosu, 1996; Nwosu, 2001). PR is now recognized around the world as a chartered profession with distinct body of knowledge, functions codes of standards and codes of ethics. In Nigeria it was chartered in 1990 by the NIPR Decree No. 16, which is now an Act of the Federal Republic of Nigeria. Based on its eclectic and inter-disciplinary nature, PR has been described as the surest route to top management in an organization (Nwosu, 2001) and has been predicted to emerge as the most popular management function of the 21st century (Baker, 2002).

The three types, forms or approaches to PR management which the environmental and other managers who want to adopt and apply its tenets, techniques, strategies and philosophy are the PR Policy Practice Approach, the PR Consciousness Approach, and the PR Practice Approach. In adopting the PR Policy Approach, the manger concerned must ensure that public relations philosophy and strategies are ingrained in every aspect of the Corporate Policies and Plans of his organization (e.g. marketing, production and personnel policies and plans), and the PR consciousness approach simply requires the manager to ensure that all categories of staff in this organization, from the lowest messenger, cleaner or gateman to the top most mangers (including the Chief Executive Officer or CEO) and Board members, are public relations conscious, PR driven, PR literate and PR orientated. The PR Practice approach requires that the manager ensures that there is a well-equipped PR department in his organization and that this department is manned by adequately trained and professionally registered PR practitioners who must work with outside PR consultants whenever necessary, and handle all the technical and professional PR functions.

Environmental Public Relations: Explanation, Strategies and Techniques

For our purpose in this chapter, we shall understand environmental public relations (EPR) as the deliberate, systematic and sustained application of public relations models, strategies and techniques and philosophy in managing all activities (e.g. campaigns and projects) aimed at protecting and preserving the environment for the human beings of today and tomorrow, as well as managing the issues and problems arising from the above activities and offering practical solutions to them. EPR as a specialized area of public relations is concerned with the result-oriented management of images, reputations, attitudes, opinions and behavioural problems, crises and issues that arise from, impinge upon or are related in one way or the other to sustainable environmental protection, preservation and sanitation.

EPR emphasizes planned activities that are supported by effective two-way communications aimed at creating adequate awareness on environmental issues and problems as well as persuading all environmental stakeholders to imbibe positive environmental behaviours or avoid destructive and deleterious environmental behaviours. EPR emphasizes relationship building and sustenance as well as partnership and participatory team work in handling all environmental issues, crises and problems. It also emphasizes persuasive approaches as opposed to coercion in managing peoples' attitudes, behaviours and crises that affect the environment. We have explained the other dimensions of EPR earlier in this book.

Environmental Public Relations (EPR) Models, Strategies and Techniques

Having explained the meaning and various dimensions of environmental public relations (EPR) again, we shall now devote the rest of the chapter to what we believe are practical models, strategies and techniques of EPR that can be employed in various contextually modified forms in managing environment projection or preservation campaigns as well as helping to reduce or eradicate the various contentious environmental issues and problems that usually metamorphose into conflicts or crises, if not properly managed. Most of the recommended EPR models or strategies are original to this chapter while a few others were adapted from the PR literature and contextually modified for the chapter.

The Triple a (AAA) Model or Strategy (Nwosu, 2003)

The triple A or AAA strategy is the first EPR strategy or model we will like to introduce and recommend in this chapter because of its basic, simple and yet practical nature. Any environmental protection, preservation or sanitation manager can easily understand it and apply it as it is or in any modified form in managing any environment, sanitation and anti-pollution campaigns and achieve the desired results. In this model or practical strategy, the first A stands for *Awareness* creation, the second A stands for *Acceptance* while the third A stands for *Adjustment* (Nwosu, 2003). How can we apply or use this EPR model or strategy. It can be used in many ways. But this model simply draws our attention to the fact that the best place to start in running or trying to manage any environmental campaign is to first and foremost create adequate *awareness* of the problems, issues or desired changes that are at the core of the environmental management effort. There seems to be at the moment too much inbreeding and wrong assumptions among the so-called environmental experts which this first A will help to reduce or remove. For example, many environmental management experts believe that most Nigerians know or should know the major problems and issues in environmental management like the *ozone layer* depletion, biodiversity, waste conversion and others.

But the truth is that many Nigerians do not know and have not bothered to know the real meanings and implications or the impact of these on their lives or the lives of the next generation of their kits and kins. Even many of the so-called educated and other elites are stark illiterate when it comes to such technical aspect of environmental management as the examples we gave above. As for the more than 60% functionally illiterate Nigerians most of who live in rural areas or urban slums, it is still difficult for them to have full grasp of the meaning and negative impacts of the common environmental issues and problems like bush burning for farming purposes, reckless cutting down of trees (deforestation), reckless defecation or disposal of human wastes by "bush methods" or into streams or rivers from where or close to where they get the water they drink, to bother themselves about such abstract environmental issues like the ozone layer depletion and others. It would also be an exercise in futility to let these Nigerians understand that there is a direct positive correlation between environmental degradation and their low poverty levels or health conditions. In fact, most illiterate rural dwellers will first believe empty fairy tales and superstitious beliefs as much better explanations of their poverty, deaths and illness than environmental problems (Nwosu, 1992).

So, in line with the first A component of our working Triple A model, we shall start any environmental problem or issues management campaign by launching a well-researched, planned and executed Awareness campaign made up of phases of actions and communications that start with and end with the target populations of the campaign in accord with modern EPR campaign management practices which abound in the PR literature, and with them participating fully (Nwosu, 1996).

It is only when the target population of any environmental management campaign is adequately aware (i.e. "knowledge-about") of the object or subject of the campaign that we can move to the second A or *Acceptance* level of the campaign team, working participatorily with the target population, will design and systematically implement series of appropriate information, education and communication (IEC) programmes, combined with various participatory activities (actions and events), over a reasonably long period (e.g. 6 months), with the objective of deeply educating the target population on the environmental problem or project at hand to the extent that they start accepting and eventually accept fully the environmental project as being in their individual, group and overall interests. This can be described as the "Deep knowledge" stage as opposed to the "knowledge About" stage of the Awareness or first A stage. This deep knowledge should lead to positive changes of attitudes and opinions in the target population (Nwosu, 2003).

The Third A in the Triple A model refers to *Adjustment.* In this last stage of the environmental management effort, members of the target population are expected to have so accepted the environmental programme or project that they are now ready to, *ceteris paribus*, to make the desired positive adjustments in their behaviours which will translate to success for the environmental manager and his team whose role at this stage will then be to catalyse, facilitate and quicken the behavioural adjustment process, making sure it remains in the right direction and does not suffer any backsliding. To achieve this purpose or objective, the environmental manager and his team should embark upon extensive communication and practical activities or events that are aimed at consolidating and evaluating the positive behavioural changes made by the target population and supporting as well as reassuring them with unshakable evidence that the environmental decisions, actions and behavioural adjustments they made are not only in their individual and group interest, but also in the society's interest, now and in the future (Nwosu, 2003).

THE STPP STRATEGY IN EPR MANAGEMENT

The STPP strategy is again another practical strategy we can employ in any attempt to use PR and PR-related strategies in managing environmental problems, issues and projects. This strategy which has its root in product marketing, social marketing and marketing public relations (MPR) is presented and strongly recommended in this chapter as the key to greater efficacy in reaching our target populations in environmental management campaigns. Its importance lies in the fact that many environmental management campaigns have been known to fail in developing countries because of the many misses and near misses resulting from inadequate, wrong or ineffective delivery of the persuasive and convincing messages that were expected to change their attitudes, opinions or behaviours in the right direction (Nwosu, 1986).

To avoid unwanted situations such as the above and so achieve maximal results in our environmental campaign efforts by hitting the right targets, we have to understand the lessons of the STPP strategy and be able to apply it correctly.

The S in the STPP strategy here refers to *Segmentation*. It calls for a careful subdivision of members of target population into manageable small groups for easier and more effective handling. This subdivision should usually be based on well-defined segmentation criteria that will ensure that as much as possible, those with similar characteristics are put in one segment, target group, subdivision or "pocket". The criteria could be based on socio-cultural, economic, political, synchographic, socio-graphic or other such information or data (Kotler, 1998).

The next step in this strategy is the T or *Targeting* stage. And for our purpose in this chapter, this involves targeting not just the segments we produced above from the target population but also specific factors like their interests, needs, wants, opinions, attitudes and behaviours that are relevant to the environmental project we are dealing with at any particular time (Nwosu, 2003).

The first P in the STPP strategy refers to *positioning*. In the context of this chapter, positioning can be described as a planned attempt to rent a space in the minds of members of the target population in an environmental management campaign, such that they will live with, sleep with, wake up with and always remember or bear in mind whatever opinions, decisions, actions, attitudes and behaviours related to the environmental programme or project in the way in which the campaign management team want them to, and how they themselves want it.

The second P in the STPP model refers to *Penetration*. It is the last step in this strategy which requires the environmental campaign manager and his team to now penetrate each segment in the target population, physically and systematically, carrying the messages, material and other things being "marketed", being advocated or being pushed or put forward for adoption by members of the target population (e.g. an environmental management innovation or equipment). In penetrating the segments of the target population, the environmental campaign managers are advised to adopt a multi-channel and multi-contact approach that will ensure maximum penetration, permanent or lasting attitudes and behavior changes on the environment.

The Ricee Model of PR and EPR

The RICEE model has its root in the original RICE model propounded by this Ikechukwu Nwosu (Nwosu, 1996). It is a model whose practical applicability has been tested and demonstrated in various contexts (Uffoh, 2002).

It is mainly a promotional model that can gainfully be employed in environmental and other PR projects for projecting specific environmental projects, programmes, issues and problems to specified target population, with the aim of influencing their awareness, knowledge, attitudes, opinions and behaviours patterns, in relations to the specific project or problem, in the direction we desire (for or against). Research and experience support the using of the model effectively as a public enlightenment and operational model that can be used in various contextually modified forms. We therefore recommend it to managers and organizations involved with tackling, promoting, projecting and "selling" any environmental innovation, idea or material to any target population. It should be seen and applied as a step-by-step and systematic strategy or model of PR, and especially EPR.

How does the RICEE model work and how can it be applied in real-life environmental management situations. First, the R in the model refers to *Research* and advises the environment PR manager and his team to start with research, environmental scanning or situation analysis. This will help them to procure all relevant data and information that will help them in coming up with the correct plans for the environmental campaign effort at hand, instead of depending on guesswork, hunches or gut-feelings. The I component in the RICEE model refers to information and require the team to design and systematically implement an information dissemination or awareness creation plan for the relevant publics" or target population (Nwosu, 1996:147; Nwosu, 1992).

The C in the RICEE model refers to Communication and reminds the environmental management team applying the EPR strategies that the information dissemination above is not sufficient unto itself because it is a one-way process. It has to be backed up with well-planned and implemented two-way communication effort. Such an effort will begin and end with the target population. This will enable the team to procure relevant, fresh or up-to-date information from them and using these to produce and implement effective two-way communication package that will ensure proper monitoring, and feedback as well as perform other relevant communication functions in the project such as the agenda setting function that helps to force the environmental issues or problem unto the public agenda for solution-oriented debates that will translate to resounding success for the environmental programme or project. The communication efforts should include proper media planning, mixing and implementation effort that integrates the modern mass media, ICTs, inter-personal, group and traditional communication media, in a manner that will ensure that the selected media for the project complement each other by bringing their individual strengths to bear on the weakness of the other media in the mix.

The first E in the RICEE model refers to *Education*. At this stage, the environmental management team is expected to use pedagogical and education methods to reinforce or strengthen the information dissemination and communication efforts they had carried out earlier. It is a consolidation strategy that will ensure lasting effects, greater imbibitions and internationalization of the desired environmental behaviours among the external publics or target populations of the environmental promotions effort. It should also be used to update, re-think, re-orientate and reinspire members of the environmental project team in the organization on various aspects of the project to help them remain focused and motivate them to keep the campaign steadily on course, as well as resharpen their knowledge and skills for the environmental job at hand. This education phase usually involves the conceptualization, production and use of various teaching aids, organizing intensive practical workshops and applying other such educational methods that will promote the acquisition of new knowledge, updating existing ones, developing of new knowledge, updating of old or existing negative attitudes and ensuring improved overall behavioural changes that will help to achieve the objectives of the environmental management projects at hand.

The second E in the RICEE model refers to Evaluation. And here, the environmental management teams are required by the model or strategy to carry out an extensive assessment of the performance of the team in every aspect of the project. Well planned and executed evaluative research is usually called for in order to precisely measure to what extent the team achieved or failed to achieve the stated objectives of the environmental project. Because data and information collected at this stage should be used as lessons that will guide future actions and decisions on the same or other projects, any evaluation effort to be carried out at this point should be conceived and implemented as a means to an end, instead of as an end itself. Popular PR and EPR evaluation efforts that could be carried out at this stage, depending on the objectives of the project, include but are not limited to the following: overall output or performance analysis, message exposure and impact analysis, gross impression assessment; awareness, acceptance adjustment and adoption/adaption assessements; attitude, opinion and behavioural change assessments and others (Center, Cutlip and Broom, 1990). Chapter Ten of this book provides deeper research insight on the RICEE model which has earlier been introduced in part one.

The Social Marketing and Societal Marketing Strategies

Two other closely related EPR strategies we shall discuss and recommend for successful implementation of environmental projects and programmes are the social marketing and societal marketing strategies. And under the social marketing strategy we shall discuss and recommend the 5Ps Social Marketing Model. Social marketing can be described as a strategy of marketing that is used for marketing or selling not-for-profit (e.g. development and environmental ideas) to specified target individuals or groups. It is related to the original profit-oriented product and service marketing strategies, but is understandably different in its approach and techniques.

Developed from the original 4Ps of marketing introduced by Mc Carthy, E.J. (1982) and popularized by others like Kotler (1998), the 5Ps model of product and social marketing is based on the New Marketing Concept's insistence that we must be customer-oriented and customer-driven in all that we do (e.g. decisions, planning and execution), in order to give maximum satisfaction to his needs and wants, create new customers easily and keep old customers or publics of our EPR-oriented environmental management campaigns who we must recognize and treat as kings, show how our environmental protection programmes or projects can satisfy their needs and wants, emphasize what benefits they would get by going along with us or accepting the ideas on the environment we are trying to "sell" to them, show them that we care (customer care), relate to them naturally and continually (customer care), and not in a touch-and-go manner, and be driven or motivated always by our deep desire to satisfy their needs, wants and interests. These done, we can be sure to get more positive actions and reactions or responses to our environmental programmes, including positive changes of people's attitudes, opinions and behaviours in relation to the environment (Nwosu, 2002:58-89).

To ensure even greater success in practical or real-life environmental projects, the tenets and techniques of the 5Ps model of social marketing must be applied systematically. In doing this, we must be guided by the fact that the first P which refers to the *"Product"* in product and service marketing, refers to the environmental ideas, habits, activities and so on that we are trying to market or demarket to the target populations or groups (the customers), or asking them to adopt or reject. We market positive environmental ideas or practices and demarket negative ones. The second P refers to *"Price"* and represents such things as the time, energy, habit change, efforts and the few sacrifices to be spent or made by the members of the target population in order to get the benefits of responding positively to the message of the environmental management programme. It is the job of the environmental team to present the benefits being marketed as much higher than whatever "price" to be "paid" by members of the target population (Nwosu, 2002:58-89).

The third P refers to the *"Place or Physical distribution"* variable in the 5Ps model of social marketing. It requires the environmental management team to develop as many appropriate channels or methods for making contacts with the target population members to make the environmental protection idea, innovation or material *available and accessible* to members of the target population, so that with little or no effort they will get whatever we are trying to "sell" to them, using the social marketing strategy. The fourth P in the Social Marketing model refers to *"Promotions"*. It demands from the environmental managers the ability to use appropriate promotional mix or marketing communication mix to ensure that the messages of the environmental campaign reach and have the desired impact on members of the target population in focus. The elements of the mix include such practices as advertising, event marketing/management, direct mail, publicity etc (non-personal methods) and face-to-face, house-to-house, telephone and others (personal methods). In modern social and other marketing, the popular approach to promotions or marketing communications is the integrated marketing communications (IMC) approach that draws elements from the non-personal and personal marketing communications families and blends them well with the other elements of the IMC like consumer insight, the Sixth Sense and Consultancy (Nwosu, 2001).

The fifth and last P of the social marketing strategy refers to *"Politics"*. It requires the environmental protection managers to ensure to take adequate consideration of the political dimensions of their efforts, factor them into their campaign plan and take appropriate actions to address whatever negative impacts these might have on the environmental campaign as well as exploit the positive potentials of these political actors and factors, for the greater success of the environmental campaign. For example, in places like the Niger Delta areas and even at the global level in which environmental issues are highly politicized, it will be naïve for the environmental management team to be ignorant of or neglect the political realities immanent in the environmental issues, problems and crises. This will be akin to walking in a minefield blindfolded and will surely spell doom or absolute failure for the environmental campaign efforts.

The *Societal Marketing Strategy,* which is sometimes confused with the social marketing strategy, is a relatively new marketing philosophy that renews the old call that marketers, like all other businessmen and women, should show real or practical respect for the Corporate Social Responsibility (CSR) theory of business. This they should do by doing such practical things as supporting or contributing to education, development projects, agriculture and other things that will add value or lend quality to their customers' lives. Interestingly, this philosophy of marketing (societal marketing) specifically demands that marketers and other business people make concrete contributions to the protection and preservation of the physical environments, especially in the societies or communities in which they do business.

So, the adoption and application of this strategy should be a natural imperative for success in environmental management that must be taken seriously by the mangers of the environment. In fact, the adoption or non-adoption of the social marketing strategy by the environmental team will determine, to a large extent, the success or failure of their environmental management projects. The target population of our environmental management campaign must therefore see us as partners who are genuinely interested and actually involved in other areas of their lives, including education, poverty eradication, health, provision of roads and other physical structures and so on. Otherwise, they will find it difficult to believe and trust us enough to make the necessary attitudinal, opinion and behavioural changes that will ensure environmental protection and preservation (Nwosu, 2004:45-56).

Issues and Crises Management Strategies

Environmental management teams that operate or hope to operate in crises-ridden areas like the Niger-Delta areas of Nigeria, where environmental problems are indeed major culprits or causes of many of the crises or conflicts, must learn the tenets of, and techniques and approaches to issues and crises management, in order to achieve any results. In fact, in such areas, environmentalists are often erroneously seen to be either part of the problems or crises, or agents of those who are responsible for the problems or crises, which the people are facing. So, they must not only be really active participants in finding solutions to these crises or problems in these areas, but must also be seen by the people to be truly interested in solving these crises, especially those related to the destruction of their environments.

In this chapter, we cannot find the space and time to go into the details of issues and crisis management. Luckily, Ikechukwu Nwosu and several others have published extensively on the application of various issues and crisis management techniques in handling public relations and other problems. (Nwosu, 1992 & 1996; Nkwocha,1999).

Environmental managers are advised to procure some of these materials and go deeply into them to expand their knowledge and skills in the area of issues and crisis management. Whatever types of crisis, the relationships between crises, issues and problems, causes of crises, crises prevention methods, specific crisis management methods or techniques research and evaluation techniques, in crisis management (e.g.

environmental scanning and SWOT and TOWs analytical techniques), and the application of the Crisis Life Cycle Model in managing issues and crises (Nwosu, 2006).

Conclusion

So far in this chapter we have painted again the unfortunate picture of the seemingly abject neglect of the world's, and especially Africa's environmental problems by the leaders, made a case for them to involve their peoples more in their environmental management plans and activities, proposed the *followers-led participatory environmental management paradigm*, pointed out that effective public relations is a major missing link in the environmental management equation or programmes, and offered and explained five major EPR and PR-related practical strategies, models and techniques for the effective application of public relations in environmental issues, problems and crises management. These are the Triple A or AAA Model or strategy, the STPP model or strategy, the RICEE model or strategy, the social marketing and societal marketing strategies, and issues and crises management strategies. Each of these strategies or models has many practice-oriented or applicatory techniques embedded or intrinsic in them, which we also explained.

We sincerely believe that if these models and strategies are systematically applied in environmental management efforts, individually or in combined forms (as appropriate), outstanding results will be registered. At least good progress will be made from what the situation is at the moment.

We like to restate, however, that the world, African and Nigerian governments have serious inputs to make in the entire process and that the aims and objectives of our environmental management efforts can be achieved. All that these governments need is the political will to do what is right in protecting the environment for the people of today and preserving it for generations yet unborn. These governments should stop paying lip services to environmental issues and problems. What is needed is practical actions, in addition to creating the enabling environments for effective environmental management and practices.

How can they take these practical actions? All the African governments need to do is to translate into action the various resolutions made in Durban South Africa between July 8-19, 2002 during which all African Presidents met to change the Organization of African Unity (OAU) to the African Union (AU), which they said will be more robust and active as well as provide a strong pan-African platform and network for launching a determinate war against poverty, health problems, bad governance underdevelopment and, of course, the many environmental problems in the continent.

At Durban too, African heads of government gave full adoption and acceptance to the New Partnership for Africa's Development (NEPAD) document, which was first produced at the 2001, and further refined in Abuja, Nigeria in October 2001. As far as this book is concerned, our prayer is that the African heads of government will truly use NEPAD as a platform to translate to action the following lofty ideas on the poor state of the environment and how to improve it, which are contained in the NEPAD document:

> It has been recognized that a healthy and productive environment is a prerequisite for the range of issues necessary to nurture this environmental base which is vast and complex, and that a systematic combination of initiatives is necessary in order to develop a coherent environmental programme. This will necessitate that choices be made, and particular issues be prioritized for initial intervention" (NEPAD, 2002).

This is good talk that is well said and we cannot agree more with its contents. But like we have been pointing out throughout this book, we need to quickly go beyond good talks to concrete actions, before it is too late for Africa, Nigeria and mother earth.

In this connection we call on the leaders and experts that met in South Africa in September 2002 during the World Summit on Sustainable Development which had environmental management as its major focus, to ensure that the summit was not just another talk shop, but a utilitarian and practical-solution-oriented summit, that would yield concrete strategies for immediate action. And we like to repeat once more that public relations and EPR-related strategies must form a significant part of these practical-solution-oriented strategies, for the many reasons we have given in this chapter.

Finally, as part of the way forward, we must ask and give practical answer to the following question: What environmental future do we want for the world, Africa, Nigeria, the Niger Delta or any other area? In answering this question we like to formulate and suggest the use of what we shall describe here as the 4Ps *Model of Futuristic Environment Issues/Problems Analysis and management.* In this our proposed model, the first P which stands for the Probable future will help us to properly identify those environmental issues that are very likely to be based purely on *Chance,* and so come up with practical strategies for handling them when they occur, without being taken unawares. The second P in the Model refers to what we shall call the Possible Future, which is self-explanatory. The possible future scenario or analysis in practical terms will help us almost in a similar way as the probable future scenario or component of the 4Ps model. But they should not be taken to be one and the same thing because, as we all should know, something (e.g. an environmental problem/occurrence) could be possible (in terms of occurring) but may not necessarily have reasonably high probability of occurring. This kind of reasoning will also help us in prioritizing environmental issues and problems for action (Nwosu, 2003).

The third P in our proposed 4Ps Model refers to Predestined Future. It can also be described as the "Act of God Future", as far as the management and analysis of environmental issues and problems are concerned. Much as these kinds of environmental issues and problems are above our human control, we should not just fold our hands and do nothing about them. This will tantamount to naivety and dangerous adoption of the unscientific and fatalistic interpretation of predestination, which many experts in philosophy do not subscribe to. If we do not do any other thing at all about the Predestined Future scenario of environmental management, we should at least come up with game plans or crisis management plans for dealing with them when and if they occur, (Nwosu, 1996).

The fourth and last P in our 4Ps Model of environmental management, stand for the *Preferable Future*. This is the desired or dream future for the environment, which we all have been and shall systematically and consistently keep working towards. It does not matter if we achieve it in its hundred percent or perfect form because it is somewhat idealistic and utopian. Also, as we noted under the Predestined P scenario above, some environmental issues and problems will indeed be acts of God that man, with all his imperfections, does not have the power to prevent. But we should not have any doubt that if we do all we should do in caring for, respecting and properly managing the environment, using or applying the strategies that now exist, we should be able to achieve a very high and comfortable level of our preferred or dream environment for the people of today and for future generations, with the help of God.

Chapter Eight

Towards An Integrated Model For Disseminating Information On Environmental Public Relations Management

Introduction

Man owes it as a duty to humanity and divinity to perpetually preserve his environment. The traditional African man and woman, no matter their political, ideological, religious or other inclinations, have always seen the environment as a divine gift. This partly explains why traditional Africans worshiped aspects of the environment, including the sun, the land, the sea, some trees and some animals.

Beyond this divine perspective, the traditional African man and woman have always placed high premium on the environment and all it offers because they see it as the key source of survival on earth. For instance, they have always been aware of the fact that they depend on the land for agricultural or farm produce, the forest for meat, the sea for fish and water and atmosphere for air.

But the problem is that much as most traditional or rural African men and women recognized and value their environment, they tend to take many of what the environment offers for granted or see them as factors which are controlled by the divine or unseen forces, Environment equilibrium to most of them has been and will always be maintained by natural forces, with or without much intervention by them (Nwosu, 2004).

But the realities of the so-called modernization or industrialization of the rural and traditional societies (Lerner, 1958) or the more current development paradigms (Rogers, 1976) make it imperative that this disposition of the traditional or rural African men and women be changed for the better. One of the major reasons for this is the fact that modern development brings with it many hazards and problems, which disrupt the original environmental equilibrium which the traditional or rural African was used to.

This chapter attempts to conceptualize and offer some theoretical and practical models and strategies for promoting awareness of environmental issues and problems in African rural communities. Our assumption is that increased awareness can lead to a positive change in their attitudes, opinions and behaviours and that it will help them to make more position contributions towards the preservation of the environment (Nwosu, 1996).

Rationale

This chapter is informed by the need for the emergence of a composite or integrated model, which can be used by those involved in the promotion of various causes, ideas, projects or programmes in diverse rural settings. At present, there are several such models, which have been applied in the developing areas of Africa, Latin America and Asia. These range from the now discredited Stimulus-Response models, the Diffusion of innovation models and the Personal Influence models to the Multi-media, Audience-Centred and Social marketing Models (Nwosu, 1991).

It will no doubt be useful if, after a close study of these models and their applications in various contexts over the years, we can come up with an integrated model that will draw from and synthesize various insights offered by these models. This will among other things, save those involved in promotion, mobilization and information dissemination in rural communities the time, money and energy spent in wading through the research and professional literature to excavate, study, understand, in bits and pieces, the many models or research insights that now exist on the subject matter. It will also provide a handy working model, which can be accessed, studied and applied in modified forms appropriate to the projects at hand.

Such an integrated model does not yet exist, probably because of the emergent nature of the various fields of study such as communication, marketing, advertising, public relations and information management from which most of them emanate. Another reason may be that the scholars and practitioners or professionals in these and other related fields have not considered it necessary and practical to work on and produce such a working model. However, we believe that such a model has become necessary and will be found useful in many ways. As Chafetz (1978) has noted, without such models or theoretical constructs "there is no science, no intellectual comprehension of the world, only a mechanical, robot-like collection of random bits and pieces of really" (p.2).

All these should serve as sufficient rationale for our attempt to suggest an integrated and conceptual model in this chapter. However, in doing this we should not lose sight of the fact that the integrated model suggested is basically a descriptive representation of the relationships among components of a system, structure or reality (Fisher, 1978). It should not be seen as a substitute for that reality.

Communication and Integrated Rural Development

To be able to understand and apply the models proposed in this chapter for promoting environmental problems and issues in rural Africa, we should be familiar with the literature on communication and integrated rural development from which many of the building blocks for the models were drawn. For instance, works, by African communication scholars such as Nwuneli and Opubor (1988); Obeng-Quaidoo (1980); Modemeka (1991) and many others, have offered recent and useful critical perspectives on the relationship between communication and rural/national development.

These African scholars emphasize such factors as the need for:

(i)   Participatory communication through community;

(ii) Grassroots and audience-oriented communication approaches;

(iii) Decentralization of information structures; and

(iv) Conceptualization of Western communication concepts, ideas, strategies or practices to make them more useful for African development milieus.

They also stress the need for development communication approaches to be related as closely as possible to the local, cultural, economic and socio-political realities of Africa. They have similarly called for the formulation of well-defined national communication policies which will, among other things, ensure balanced information flow within national entities and overcome the present rural-urban information imbalance and socio-economic inequities in African countries.

In addition, these African communication scholars seem to have a strong faith in the role of communication, either as an independent or as an intervening variable, in the development process. As an intervening variable, they tend to emphasize the influence of communication media at the awareness level of information processing in development. As an independent variable, they emphasize the indirect role of the mass media in attitude and behavioural change along with or through a nexus of other inter-personal, group and traditional media of communication in the society concerned (Nwosu, 1990).

All these are in accord with McGuire's (1989) insistence that in development communication efforts (as in promoting environmental issues and problems), greater emphasis should be placed on harnessing the multiplication potential of the mass media, which makes them teachers, informers and awareness creators. This means that the mass media must be properly integrated with personal, interpersonal or group communication methods for message multiplication effect and for the possible change of attitudes, behaviours or opinions.

In an indepth analysis of the problem of development communication, Mody (1991) generally endorses the above ideas. But she warns that, because audiences are not passive internalizers of media messages, whatever the media present are modified by the prevailing cultural and other variables, resulting in partial acceptance, reinterpretation and sometimes outright rejection of mass-mediated messages. She recommends what she describes as a Systematic Audience Dialogue-based Approach to any type of development communication efforts, including those aimed at promoting desirable environmental behaviours. This approach stresses the need for horizontal communication within and between groups and vertical or people-to-planner information flow on needs, priorities and preferred modes of meeting those needs. It also incorporates or makes allowance for top down or planner-to-people information flows in response to community needs. Any serious and well-planned programmes for promoting environmental issues and problems in rural Africa must be guided by the research insights reviewed above in order to register any appreciable success.

Knowledge of Environmental Issues and Problems in Rural African Communities

Another critical requirement for the proper understanding and application of the environmental promotion and information dissemination models proposed in this chapter is a thorough knowledge of and familiarity with peculiar environmental issues and problems in African rural communities. In this section, we shall, therefore, offer a summary of these issues and problems which we have described or discussed in some detail in other chapters of this book.

An important point of departure is to note that there are some clear differences between the main environmental issues in Africa and those in the developed societies. There are also significant differences between the environmental issues and problems in rural and urban Africa. This is understandable because the environmental problems of any area or region are closely related to its peculiar geographical, demographic, technological and other kinds of development, as well as other localized factors, which include peculiar cultural practices and beliefs. However, there are global environmental issues and problems which rural communities in Africa share with the rest of the world by virtue of belonging to the planet earth. Nevertheless, our concern in this chapter is related more to the peculiar rural African environmental issues. This is not to belittle the importance of global environmental issues and problems. It is just a way of saying that, in promotion and information campaigns in which changes in deep-seated attitudes or behaviours are the ultimate goals. It may be better to start from the known to the unknown or from the familiar to the unfamiliar.

The environmental issues and problems we need to be familiar with in the context of this chapter include those mainly published by the United Nations Environmental Programme (UNEP) in the African context. Among the most critical problems are water pollution and spread of water-borne diseases like diarrhea, river blindness and malaria in rural Africa. It is estimated, for instance, that 2 million people in the world die of malaria every year, while some 100 million people are infected annually. It is significant to note that most of the people involved in the above-cited statistics reside in rural and urban tropical Africa. Environmental literacy and proper management of the environment can help change this situation.

Another serious environmental issue in rural communities in Africa is the use of wood, animal dung and agricultural wastes as fuel for cooking and other purposes – something, which poses serious environmental health hazards. Also, the drive towards rural industrialization in the past two decades has brought about the existence of factories in some parts of rural

Africa. These factories produce dangerous toxic wastes and pollutants, which are detrimental to air, farm-lands, animals and those who reside in those areas, as we had explained in some detail in earlier chapters of this book.

The pollution of the African rural environment is made worse by certain harmful practices among the rural dwellers. Among these practices are unhealthy human, animal and other waste disposal systems and indiscriminate bush burning. In many rural African communities, the pit latrine and the bucket system of human excrement disposal are still in vogue and exist side by side with the "bush method" which refers to using the open bush as toilets. All these practices are undoubtedly unhygienic as they do not only pollute the natural land environment but also cause the spread of infectious diseases through air and water. Although this situation is related mainly to the socio-economic development of these rural African communities, well-organized environmental campaigns can help to positively change the situation by inculcating good health and environmental practices among the rural dwellers, at least until they get modern sewage and waste disposal systems (Nwosu, 2004).

Perhaps the most devastating of the environmental problems in African rural communities are those related to natural disasters like floods, erosion and desert encroachment which we discussed earlier. While many African governments and international agencies spend a considerable amount of financial resources every year to combat these natural disasters, it is equally necessary that those who live in the rural areas are educated through well-organized promotion and information campaigns on how to prevent or reduce the occurrence of these natural environmental disasters. Among other things, they should be helped through these promotional efforts to overcome the harmful habits, beliefs and practices like bush burning, wild-life destruction and deforestation which facilitate the occurrence of these natural environmental disasters. Promotion and information campaigns should also teach them positive environmental practices like tree-planting efforts for erosion control and afforestation or forestry development (Nwosu, 2004).

The need for these environmental protection campaigns in rural Africa is heightened by the fact that Africa lags behind most regions of the world in bringing about environmental awareness among its peoples, especially those living in the rural areas. As walker (1992) puts it, "Africans have numerous environmental problems with a rapidly expanding human population and there is an urgent need to get people to recognize that it is in their best interest to be environmentally literate.

But before we can create literacy, awareness or sensitivity about the environment among rural Africans through careful application of promotion and information strategies, we have to know them intimately or well enough. This is in line with the audience participation-based promotion/information approach we adopted in this chapter.

Knowing the Rural Africans

Many programmes or projects, including promotion and information campaigns, aimed at people in rural communities in Africa often fail mainly because the planners and implementers of such programmes have inadequate knowledge of the ruralites, their cultural backgrounds, needs, likes, dislikes and characteristics. Thus, often the programme planners and implementers fail to carry the ruralites along or do not even reach them because their efforts are shot below or above the heads of the ruralites or are not related to their interests and realities. In this section, we provide a description of portrait of the typical rural African, which should be used mainly as a guide to further understanding of the specific members of the rural community who may be the focus of any particular environmental protection or preservation efforts (Nwosu, 1996).

One factor to be noted is that, even though rural Africans are normal human beings who share the same human needs and characteristics with other human beings elsewhere in the world, they have many peculiarities, beliefs, customs, experiences and varied cultural, political, educational and economic variables which make them different in many ways. Their environment is also different. To effectively reach them, raise their awareness and probably change their attitudes, opinions and behaviours in any desired direction, we must not only be familiar with their peculiarities or characteristics, but must work with them closely in order to carry them along for lasting results. These rural Africans constitute about 75-80% of the population in most African countries. So, whatever investments are made in trying to help them, remain easily defensible, justifiable and utilitarian (Nwosu, 1990).

Among the peculiar characteristics of the African rural dwellers is that, although they may speak some English, French or other foreign languages, they are not black English or French people and so should not be perceived or treated as such in any promotion or information dissemination effort for environmental protection. They are attached strongly to their traditional cultures or ways of life and have better understanding of messages communicated to them in their local languages in their pure form with idioms, proverbs and figures of speech. The rural African is brought up and lives in an environment where hundreds of languages and dialects, as opposed to one dominant language, are spoken (Nwosu, 1990).

The rural African person is brought up and lives in an environment where, in spite of the incursion of Christianity and Islam, many traditional religious practices and beliefs still thrive. The rural African may possess or have access to watches and clocks, but he is still affected by what Mbithi (1971) and others describe as the African concept of time. As Obeng-Quaidoo (1986) observes, the rural African's attitudes, opinions and behaviours are affected in many ways by the following core value boundaries of the African culture:

1.   The role of the supreme God/Allah and lesser gods in his daily life;

2.   The African concept of time;

3.   The African concept of work and its relationships to how he perceive his relationship with nature; and

4.   The non-individuality of the African and how this affects his worldview or cosmology.

In short, the rural African lives in an environment in which traditionalism and modernism coexist but which is largely underdeveloped. All these affect their life-styles, relationships and level of literacy. If we are serious about reaching them and working with them to combat environmental problems in Africa, we must know and respect them. We must also recognize that the above-mentioned factors constitute part of the rural African socio-economic, cultural, political and educational realities and reflect them in our planning, packaging and execution of environmental promotion and information campaigns (Nwosu, 1993).

Having provided the necessary background, we can now suggest specific strategies and models for promoting or disseminating information on environmental issues and problems in rural Africa.

In this section, we shall describe, explain and suggest some specific models or strategies for disseminating information on environmental issues and problems. We shall also use the models as building blocks for a composite or integrated model for promoting information dissemination on environmental issues and problems.

The Replace Model

The first model is what we have described as the REPLACE Model. REPLACE for our purpose describes the key components of the model. But it is also a concept which conveys the message that there is need for us to replace many of the old models, paradigms or approaches which we have been using to promote environmental issues and problems in rural Africa with new ones which are more related to the realities of rural communities in Africa.

Figure 8.1: **The Replace or Ante-Action-Afterwards Model**

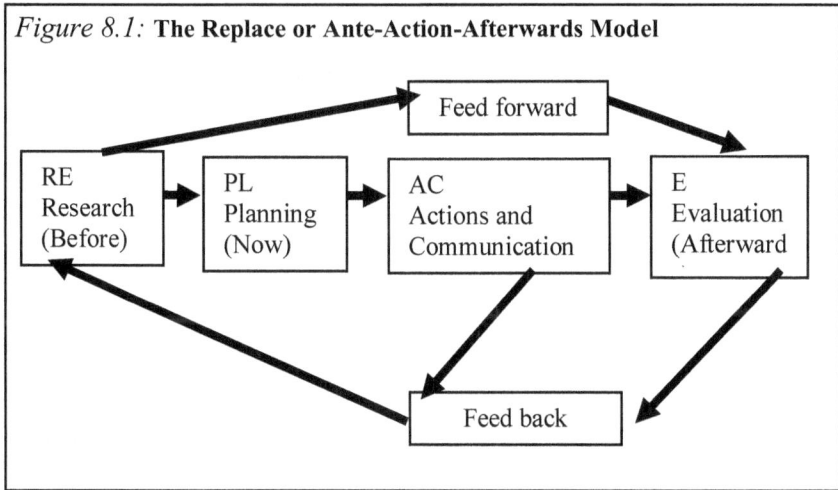

Source: Nwosu, Ikechukwu E. (1993) *in Media and Environment in Africa.*

The model essentially comprises research, planning and evaluation and emphasizes the need for all environmental promotion campaigns in rural Africa to be accompanied with adequate research, planning, on-project and post-project evaluations. In the model, which can also be described as the Ante-present-Action-Evaluation model, R refers to Research, E Evaluation, PL refers to Planning, AC refers to Actions and Communications, and E refers to Evaluation. It stresses the need for those involved in environmental information campaigns in rural Africa to:

(i)   Collect data that will adequately describe the status ante (Before) and status quo (Now) well;

(ii)   Use these facts to design their actions and the messages (Action) to be transmitted to the target audiences in rural Africa;

(iii)   Be able to measure or evaluate the results of their activities intermittently and at the end of the campaign (Afterwards).

The REPLACE model which has in-built feedback and feedforward mechanisms is represented in Figure 8.1,

As Figure 8.1 shows, the model can help the knowledge of what was, what is or what exists now in rural Africa to determine which appropriate actions and communication activities should be used to achieve environmental promotions objectives. It can also help in assessing the extent to which the objectives have been achieved over specific time spans.

Audience Participation Model

The next model can be described as the Audience Participation or Audience Focus model. It stresses the need for the experts in environmental information to endeavour to start and end their activities with the target audience (the rural Africans). This model directly focuses on the audience and is especially important and useful to message design, packaging and dissemination.

The model enables us to apply the kinds of promotion campaign techniques recommended by Mody (1991). These include learning everything possible about the subject-matter of the campaign which in the context of this chapter is environmental problems, analysing the lifestyles and communication preferences of the audience, and assessing audience needs in relation to the campaign or promotion effort. The techniques also include formulating specific measurable goals, selecting appropriate media or communication mix, formulating and applying creative-persuasive strategy, and preparing messages within well-defined specifications. Also included are the techniques of pre-testing the messages, modifying them before final message or media productions, transmitting the messages through the appropriate channels, monitoring and analyzing audience exposure to the messages, and collecting data for evaluating the impact of the messages on the audience. This Audience Participation model can be diagrammatically represented as shown in Figure 8.2.

The 5Ps Social Marketing Model

Another model that we commend for the dissemination of information on environmental issues and problems in rural Africa is the 5Ps of social marketing. An expansion and off-shoot of the 4Ps of social marketing which has been found to be useful in many information campaign efforts in Africa (Nwosu, 1990); this model emphasizes the need for those involved in promoting environmental awareness to pay adequate attention to the political imperatives of the social marketing process. This is because there is hardly any decision that can be taken in campaign effort without paying attention to its political implications or considering the political power structure in the African country concerned.

Popularized by Kotler (1977) and others, the original 4Ps social marketing model is similar to the Audience Participation model because it is a philosophy of business which stresses that the satisfaction of the customers or audience's wants and needs is the main social and economic justification for the existence of any business organization. The original 4Ps of social marketing has been expanded to accommodate non-profit making or non-business agencies or organization, like environment protection or preservation agencies, in which financial profit is not the ultimate organizational goal.

The main message of the 5Ps of social marketing proposed here is that managers of environmental campaigns should, in addition to being politically conscious, be able to perform like aggressive marketers and respect marketing imperatives.

Figure 8.2: The Audience Participation Model

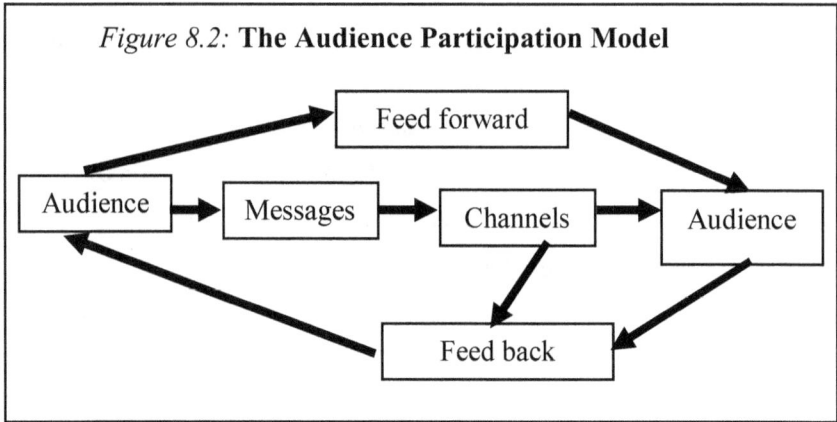

Source: Nwosu, Ikechukwu E. (1993) *in Media and Environment in Africa.*

Figure 8.3: The 5Ps Model of Social Marketing

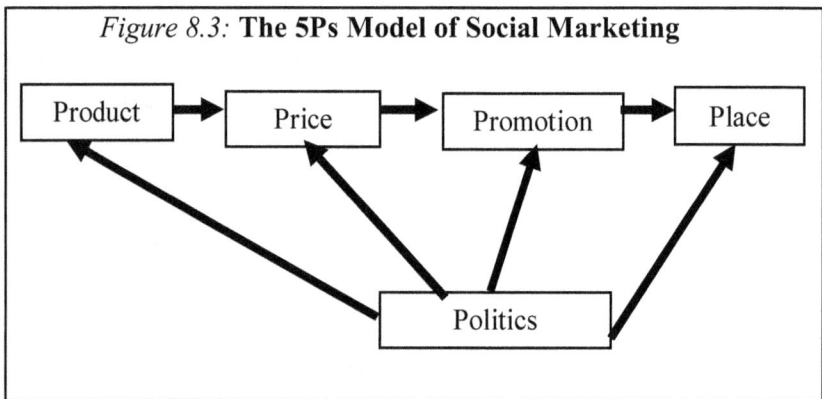

Source: Nwosu, Ikechukwu E. (1993) *in Media and Environment in Africa.*

This is because, even though they are not marketing products or services for commercial or profit-making purposes, they are marketing environmental ideas to well-defined target audiences who have every right to accept these ideas or reject them, depending on so many factors.

In the model, the first **P** refers to the product, service or idea to be sold to the people through communication and persuasion; and the second **P** refers to the price to be paid by the people or, in the context of the chapter, whatever sacrifices they have to make in accepting the proposed new environmental idea or practice. The third **P** refers to the promotional efforts, which must be carried out to ensure the acceptance of the idea or practice being proposed. These efforts include advertising, publicity, public relations and other promotional methods like house-to-house calls, face-to-face and group communication techniques. The fourth **P** refers to the place or physical distribution channels available to the environmental campaign manager for getting messages about the environmental ideas or practices to the target audiences. The fifth **P** refers to political considerations, which we have described above and which must be respected by the environmental promotions managers. This model is graphically presented in figure 8.3. It was also explained from different angles of chapter Seven of this book.

The Decentralized and Inter-Sectoral Model

The final model proposed in this chapter for the effective promotion of environmental issues and problems in rural Africa is the Decentralized and Inter-sectoral Model. This model can be applied in various forms in the design, planning, and implementation of promotional messages on environmental issues in the rural settings. It stresses the need to decentralize the patterns or flow of environmental information in African countries. Its inter-sectoral component stresses the need to realize that, since the environmental sector is a system, it will be an error to plan promotion campaigns in isolation or without adequate linkage with these other sectors or sub-sectors.

The model is diagrammatically represented in Figure 8.4.

As Figure 8.4 shows, at the hub of the proposed decentralized and inter-sectoral model are the rural dwellers who are the centre-piece or director target of the campaigns on environment, immediately surrounding them are the various communication media. The outermost ring of the model constitutes the environment of both the media or information sources and the rural dwellers. These sub-sectors range from the political, educational and socio-cultural to the agricultural, information, environmental and health sub-sectors which directly or indirectly affect the flow of information on environmental issues and problems in rural communities.

The model will be useful in decisions about financial appropriation and budgeting, which are central to the planning and implementation of all rural information flow and promotion projects. For example, the model will help us to demonstrate that ideas, activities, material and similar resources from other sub-sectors often relate to or affect environmental practices or situations and must be budgeted for.

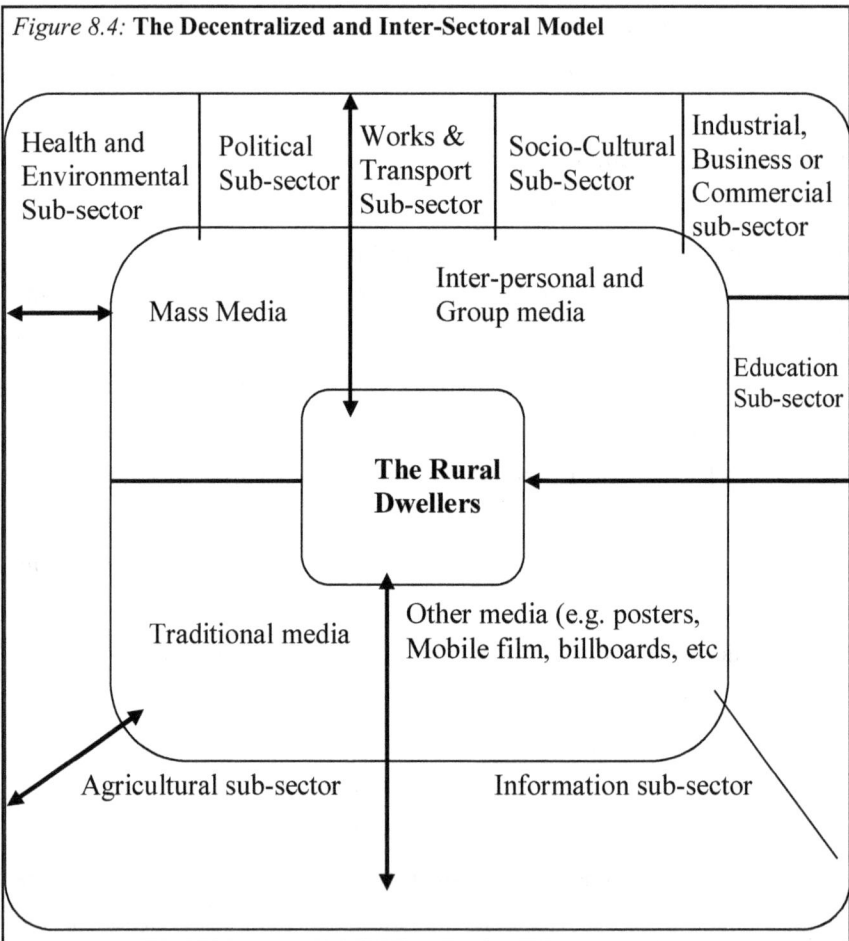

Figure 8.4: **The Decentralized and Inter-Sectoral Model**

| Health and Environmental Sub-sector | Political Sub-sector | Works & Transport Sub-sector | Socio-Cultural Sub-Sector | Industrial, Business or Commercial sub-sector |

Mass Media — Inter-personal and Group media

Education Sub-sector

The Rural Dwellers

Traditional media — Other media (e.g. posters, Mobile film, billboards, etc

Agricultural sub-sector — Information sub-sector

Source*:* Nwosu, Ikechukwu E. (1993) *in Media and Environment in Africa.*

The model can be used to demonstrate that the environmental information needs of those who live in the rural communities are closely related to and cannot be divorced from their needs in the other sub-sectors. Indeed, it is the abject neglect of these inter-sectoral interlinkages, which is largely responsible for the colossal failure of most single issue information or promotional campaigns in Africa.

The proposed decentralized and inter-sectoral model can also be used as a research model in the study of environmental issues and problems in the dissemination of information on the environment. It can help the researcher to be more holistic, balanced and sufficiently extensive in the definition of research problems on the environment, conceptualization, data collection and analysis, discussion and conclusions.

Towards an Integrated Model

A close look at each of the four models shows that none of them can stand alone as a complete conceptual and practical model for promoting information on environmental issues and problems in the rural setting. In other words, none of the four models can be applied independently in all aspects of an environmental promotion campaign, which includes campaign conceptualization, research, planning and budgeting, execution or implementation, communication and evaluation. It is therefore, necessary to use them as building blocks to design an integrated model of environmental issues and problems promotion. Such a comprehensive and integrated model would have utilitarian value in both scholarly and professional circles.

The diagram in Figure 8.5 represents the proposed integrated model.

As Figure 8.5 shows the 5Ps model, the REPLACE model, the Audience participation model and the Decentralized and Inter-sectoral model have been integrated to produce what we consider to be a composite and relatively comprehensive model. This model could serve as a theoretical and practical guide for all those involved in the promotion of awareness about environmental issues and problems in rural settings in Africa.

Figure 8.5: **Integrated Model of Environmental Information Dissemination**

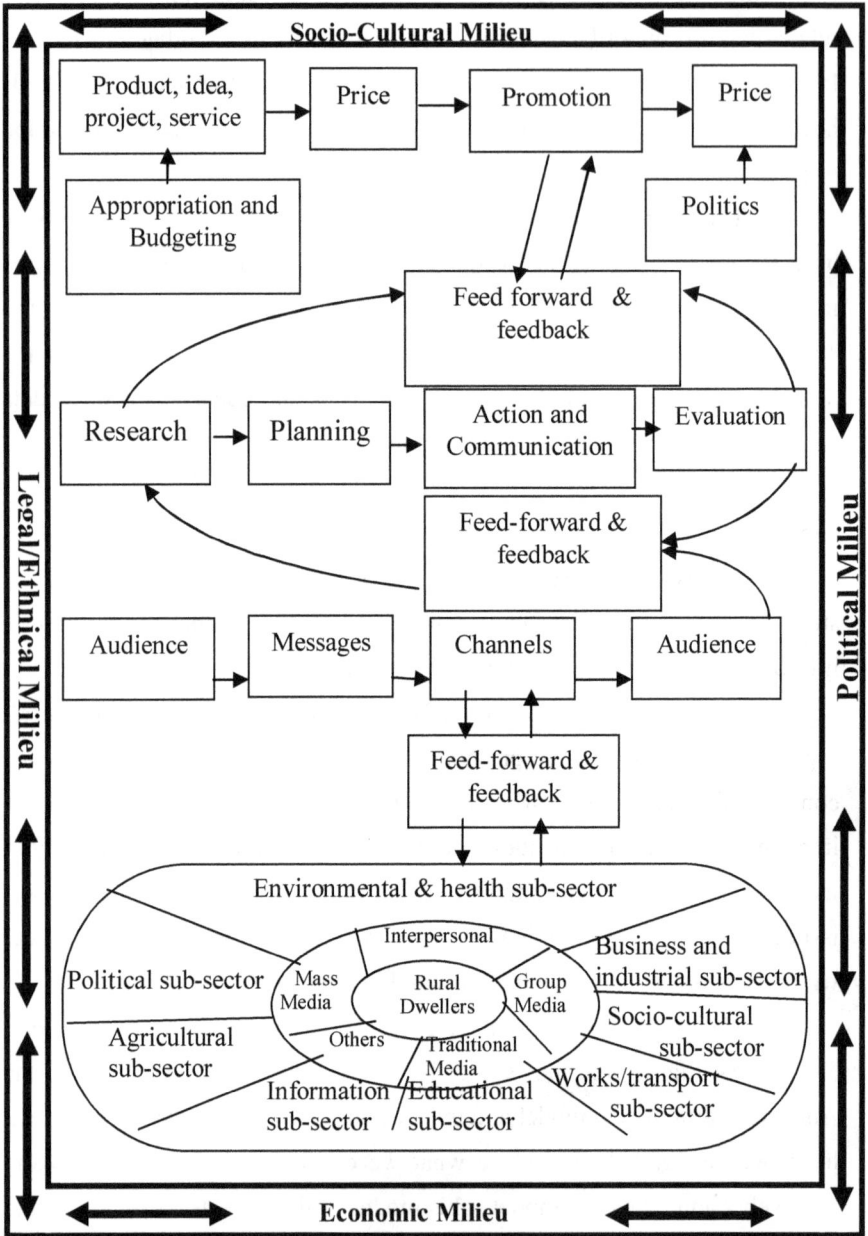

Source: Nwosu, Ikechukwu E. (1993) in *Media and Environment in Africa*

Conclusion

This chapter has attempted to apply the integrated marketing communications approach (Scissors, 1991; Okigbo, 1992; Nwosu, 2001; to the conceptualization, design, explanation and presentation of strategies and models for the promotion of information on environmental issues and problems in rural Africa. The chapter has drawn and applied ideas from five related disciplines, namely communications, advertising, sales promotions, public relations and direct marketing (Nwosu, 2001).

The guiding principle in this effort has been to create the right theoretical and practical framework for the effective delivery of information or messages on environmental issues and problems in Africa.

Chapter Nine

Environmental Impact Assessment (EIA) And Its Major Methodological Approaches

Introduction

Environmental Public Relations (EPR) Managers, as we have already stated, should have reasonable knowledge of various aspects of environmental management. While it is not possible for us to handle or treat all areas of environmental management in a single text such as this, there are certain areas that must be touched upon because of their great importance. Environmental Impact Assessment (EIA) is definitely one of those important areas.

In this chapter, we shall briefly introduce the environmental PR manager to the principles and methods of EIA, which we consider one of the most important areas. We will use this as a takeoff point to now zero in on the major methodological approaches in EIA. This is because we believe that these methodological approaches are the "Heart" of the conceptualization, planning and execution of any EIA. And so, anybody who has full knowledge of them will be assumed to have reasonable knowledge of the entire EIA.

As we noted earlier, the environmental PR Manager only needs to have a reasonable knowledge in this area to be able to work with other environmental experts in the fields brining in his knowledge and skill in his specialized area – Public Relations. Environmental management is after all, a multi-disciplinary and partnership-based practice or discipline.

Meaning, Definition and Explanation of EIA

It is man's effort to avert or at least seriously minimize the hazards caused by his unsustainable exploitation of the environment that led to the emergence of the concept of Environmental Impact Assessment. This concept ;however, first found expression with the passage in the United States of America of the National Environmental Policy Act (NEPA).

Environmental Impact Assessment according to Essagha et al, (1999:12), is an operational instrument that seeks to initiate, identify, collect, assess, establish, document and store the current and future interactions, interrelationships and systems between a proposed project and its environment. And man can always strike balance between economic development and ecological well-being. More still, the United Nations Environmental Programme (UNEP) defines EIA as a process that identifies, predicts and describes in appropriate terms the pros and cons (penalties and benefits) of a proposed development project. Furthermore, Smith (1993), views EIA as a process of environmental planning that provides a basis for resource management to achieve the goal of sustainability.

However, we must note the slight difference between environmental impact assessment (EIA) and Environmental Evaluation (EE). Whereas EIA is carried out for a proposed project that is environmentally significant, environmental evaluation is usually carried out for an already existing project to ascertain the environmental effects of such projects in order to articulate strategies for remedying or ameliorating such effects.

Goals and Objectives of EIA

The overall purpose of EIA is to enthrone a sustainable development process whereby environmental considerations are effectively integrated into the developmental processes (Esegha and Adebe, 1999:12). However, the concept of sustainability originated with the 1980 World Conservation Strategy of the International Union for the Consideration of Nature and Natural Resources (IUCN). The IUCN advanced sustainability as a strategic approach to the integration of conservation and development consistent with the objectives of:

- Ecosystem Maintenance
- The preservation of genetic diversity
- Sustainable utilization of resources (Smith, 1993:2).

EIA is meant to protect the bio-geophysical environment. The environment as defined by the European Union is the combination of elements whose complex interrelationships make up the settings, the surroundings and the conditions of life of the individual and of society, as they are or as they are felt. Thus, EIA provides a kind of framework for stocktaking within which environmental capacities can be ascertained and measured, risks analyzed,; possibilities determined and sustainable pathways defined for action (UNEP, 1988).

On another note, EIA provides decision-makers with environmental analysis so that decisions can be made based on as nearly complete information as possible. EIA also is a source of information on a proposed project to the public.

More-still, EIA evaluates, assesses and presents intangible, non-quantifiable effects, not adequately addressed by traditional appraisal technique and technical reports.

EIA also seeks to formalize the consideration of alternatives to a proposal being considered so as to ensure that the least environmental harmful means of achieving the given objective is chosen (World Bank, 1978).

Yet on another note, EIA strives, to improve the design of new development and safeguard the environment through the introduction of mitigation and avoidance measures. It also creates opportunity for consultation with the public and other interested parties (Onyeador et al, 1999:3). Some environmentalist simply view EIA as a means of avoiding or rather checking occurrence of such disasters caused by human activities on the environment.

From another perspective however, proper execution of EIA pays off in saving the money which would have been used for clean ups, settling litigations and monetary compensations. Proper EIA also quickens the approval for a proposed project. On the same note, adequate EIA ensures that there would not be any such issues as mentioned above in the later stage or date in the lifetime of a particular project and if there is, it will be very minimal. Furthermore, EIA aids in choosing best project designs and location, which in turn, will result in maximization of benefits and minimization of harmful effects.

Moreso, when public participation aspect of EIA is properly carried out, antagonisms by the public (especially local community), are eliminated and their goodwill are elicited. Thus, we conclude by observing that adequate performance or execution of EIA tantamount to compliance with the agreed environmental standards, which in the long run guarantees sustainable environment.

Basic EIA Procedures

In this subsection, we would briefly acquaint the Environment Public Relations Manager with the best processes and procedures of EIA. Firstly, we note that not all proposed projects need environmental Impact assessment. For a project to need that an EIA be carried out on it, it must be environmentally "significant". This simply means that it must be seen to have a substantial impact on the environment where it is to be cited. This process of determining which proposed project(s) are environmentally significant and therefore deserving an EIA is called screening. The modalities and criteria for this screening is spelt out in the environmental guidelines, rules, regulations and laws of a particular nation, state or local government in question. This means in effect that the same project which is seen to be environmentally significant in Geneva and therefore deserving an EIA may not be seen to deserve an EIA in Enugu State of Nigeria because of what the laws and rules say. This underscores the importance of knowing the environmental guidelines, rules, regulations and laws governing the zone of the proposed project.

Then, when a project has been certified to need an EIA, an environment scooping is carried out. This scooping succinctly explained, means determining those environmental variables, elements and factors of the said environment on which the impact assessment is to be carried out. Examples of such variables include; Aquatic flora and fauna, a particular endangered animal specie etc.

This is followed by proper description or characterization of the environment. This according to Essagha et al (1999), involves the process of identifying the various components of the environments, the different elements in each component, as well as the description of the environmental items that are important and necessary for the prediction and assessment of the impacts of the proposed project on the environment.

After Environmental "characterization", impact prediction and assessment is carried out, and this provides a platform for evolvement of alternatives as regards the proposed action. These alternatives equally include the no-project alternative. Hence the best alternative is selected depending on the outcome of the previous steps, more especially the prediction and assessment carried out.

Once all the above procedures have been accomplished and an alternative chosen, the Environmental Impact Statement (EIS) is prepared. Firstly the draft EIS is prepared, this draft EIS is usually subject to comments and reviews by all the stakeholders in the said environment. It is basically on the basis of this draft EIS that public participation is carried out. In the end, the final EIS is prepared and submitted to the appropriate authority for approval before the project is allowed to take off.

Major Methodological Approaches to EIA

We would in this subsection discuss the six most acceptable EIA methodological approaches. They include: Adhoc; Checklist; Interaction matrices; Overlay Mapping; Simulation Modelling; Network. These methodologies are the procedures, and structured techniques for performing the task of environmental assessment. We equally note that different methodologies could be used at various stages of an EIA according as it is most effective to achieve the required result.

Adhoc Method

This is the oldest EIA methodology. This method generally gives areas of possible impact of the proposed project. It does not give details of specific components or elements of the environment that are to be affected. Example, it can just indicate that aquatic life would be affected without details of which of the aquatic fauna and flora are meant.

As the name suggests, it entails putting together professionals to give a quick analysis of possible impact of a proposed project. The approach usually adopted here is that of considering each environmental section or area and indicating the nature of the impact upon it.

Checklist

Checklist equally was among the earliest and simplest methodologies developed for EIA. Checklists are standard lists of features, which may be affected by a project. They present the simplest form of assessing project impacts. They aim to promote thinking about impacts, providing a concise summary of the effects of proposals, identifying factors and the trade-offs between alternatives. Checklist transcends mere indication of possible impact areas as in adhoc to providing the components/elements that would be affected. It equally assesses character or nature of the impact.

There are several formats for checklists, however, about five types of checklist, have been endorsed by several authors. These include: Simple checklist Descriptive checklist; Scaling/Ranking checklists; Multi-tribute utility theory/weigh-ting scaling checklist; Questionnaire checklists.

Simple checklists give only the lists of potential environmental variable/factors or elements that would be affected by a particular project to be cited in an area. It does not offer any guidelines or information as regards how the various factors are to be measured or on specific data needs or how to assess significance of impacts or ways of measuring environmental change.

Descriptive checklist unlike simple checklists has measurements and predictive techniques for each variant/element included in the checklist to provide guidance on the assessment of the impact identified by the listing. Descriptive checklists are usually used for water resources projects, transportation project, military activities, land development projects, in coasted area.

On the other hand, in scaling checklists, criteria for evaluation are incorporated into the listing. This usually comes in the form of a subjective rating or scaling. It is designed as a worksheet with space to indicate the relative significance of each impact. The ranking, as it were, involves the arrangement of alternatives from best to worse, considering their potential impacts on identified environmental factors. This type of checklist would be suitable for comparative evaluation of project alternatives. Hence, it will be of great help in the selection of project alternatives.

Furthermore, multi-tribute utility theory could be seen as an extension of scaling checklists, a subjective measure of relative merit (a utility function) derived from each of the environmental parameters listed in the checklists. The impacts of the project alternatives are then compared on the basis of these derived values. Though, the multi-purpose utility approach is complex, it does however, incorporate an explicit consideration of probability and sensibility analysis.

Another type of checklist is a questionnaire checklist. This consists of questions relating to the impacts of a proposed project. Such questions are usually listed in categories. For example, community health, water pollution etc. The questions are answered with: Yes, No and Unknown, where Yes = an Impact; No =No Impact; Unknown = Uncertainty. Usually when the answer is yes or unknown, further investigation is needed. As it is, this method has proved to be most helpful to non-professionals.

Interaction Matrices

This is usually, a diagram consisting of vertical and horizontal axis with one set of either environmental factors or project activities being at either side of the diagram respectively. Then interaction between these sets of factors is recorded at their point of intersection. This interaction could indicate viz; positive impact; No Impact; negative impact or uncertainty of any possible impact.

Major types of interaction matrices include the one developed by Leopold and Co. otherwise called Leopold Matrix, Peterson Matrix and Component Interaction Matrix. In order to further get a vivid picture of what interaction matrix entails, we note that Leopold Matrix for example has lists of 100 possible interaction cells. However, not all these cells are made use of for every particular project. The ones to be used depend on the project and environmental factors or variables in the area the proposed project is to be cited.

Generally, interaction matrices have been proved to be valuable in impact identification and presentation of comparative information for different project alternatives. It has even been more useful in developing countries as a means to providing initial environmental examinations and quick impact assessments.

Overlay Mapping

Though, some authors trace the use of this method as far back as 1913, but more vividly it is seen to be pioneered by McHarg (1968). In this method, a superimposition of transparent sheets showing the sets of features of the environment before the impact according to quality is done on a base map of the area in question. The expression of this value classes is shown: Dark shading connotes high impact; medium shading implies medium impacts while low shading entails to impact. Thus, impacts are shown through the density of shading, which builds up as each attribute is overlaid.

Overlay mapping presents the environmental stakeholders with simple but powerful visual representation of possible impact of a proposed project. Hence showing where opportunities for the said project are most environmentally sustainable.

However, the limitations of this approach bring on the fact that only as little as between 7 and 12 impacts could be successfully overlaid. Morestill, there is neither distinction between primary, secondary or tertiary impacts nor is there information as regards duration of impact.

Networks

Networks are an environmental methodology designed to give a directional link between the primary and secondary or even tertiary environmental impact (as the case may be) of a proposed project. We note that the strength of networks rests in their ability to identify and guide analysis to indirect impacts, which may arise from a project (Smith, 1993:23).

The usual diagrammatic expression of this methodology makes it to be easily understood and appreciated by publics who may not be professionals. However, the methods do not offer impact evaluation but just impact identification. It is equally costly and time consuming.

Simulation Modelling

According to Onyador et al, simulation modeling or otherwise called expert systems involve the interaction between small core group of EIA experts, a bigger group of interested parties including decision makers, single discipline experts and computer- modelling experts. The computer specialists convert the output from the interaction into Environmental System Model. The model can now be used to identify impacts in future cases.

This method entails the use of mathematical and other sciences in the modelling of environmental systems. We note that this methodology is mostly useful in early EIA stage as it helps in the conceptualization of the impact assessment study. Its main advantage is that it can be used even with few data. However, it is very expensive and needs a cream of trained experts to produce a desirable result. An example of this simulation modelling for EIA is the Adaptive Environmental Assessment and Management approach (AEAM). This model has been proven to have been based on the works of Holling.

Conclusion

We believe that with this expose on the principles, procedures and methodologies of EIA done in this chapter, we have been able to acquaint the Public Relations Manager with the basic knowledge of EIA and he now has the right footing to work effectively among other environmental experts in tackling environmental issues and problems, especially in the very technical area of EIA.

However, we would not conclude this chapter without nothing that what really determines the particular methodology to use include, among other things; the kind, nature and scale of the proposed project; the proposed location of the project, the fund available; the guidelines or legislations of a particular area; the available skill; the effectiveness and reliability of a particular methodology in a particular circumstance.

PART FIVE

****

DEMONSTRATIVE/ ILLUSTRATIVE RESEARCHES AND CASE STUDIES ON ENVIRONMENTAL PUBLIC RELATIONS (EPR)

****

Chapter Ten

A Sample Survey Case Study On The Ricee Model And Environmental Public Relations

Introduction

According to the Oxford English Dictionary (1978:108), degradation or pollution entails the action of or condition of defilement, uncleanness or impurity caused by contamination. M.W. Holdgate (1993:3) maintained that pollution is the introduction by man into the environmental of substances or energy liable to cause hazards to human health, harm to living resource as ecological damage, or interference with legitimate uses of the environment. However, for the purpose of this book, we shall adopt a more pragmatic and functional definition which defines environmental degradation, otherwise known as environmental pollution, as any addition to air, water, soil or food that threatens the health, survival or activities of human or other living organism (Miller 1998:16). The particular chemical or form of energy that causes such harm is called pollutant. The pollutants are mostly solid, liquid or gaseous products or wastes produced when a resource is extracted, processed, made into products or used. Equally, it can take the form of unwanted energy emissions, such as excessive heat, noise or radiation (Miller, 1998:16). There is a great link between awareness or knowledge and action. Thus, people do not usually act for or against a cause unless they have been properly enlightened, educated and perhaps convinced of the grievous consequence of non-action with regards to that particular phenomenon. Hence, the public relations RICEE model designed by Prof. Ikechukwu E. Nwosu (1996), is assessed in this chapter to verify its applicability as a Public Relations Strategy for tackling the problem of ignorance as regards environmental pollution or degradation in Enugu metropolis.

The public relations RICEE model is a suggested operational model for the application of public relations techniques and strategies to the management and promotion of environmental issues and problems. It is recommended to public relations managers and suggested that they modify it contextually and appropriately from issue to issue, problem-to-problem, company-to-company, community-to-community... as may be required by the environmental issue management problem at hand (Nwosu, 1996:146). More still, the RICEE model deals with the topical issue of environmental public relations or how public relations can be applied in managing or controlling environmental issues and problems. In the model as we noted earlier,

- R stands for Research
- I stands for Information
- C stands for Communication
- E stands for Educational and Execution, and
- E stands for Evaluation (Nwosu, 1996:10)

Thus, the PR RICEE model is more or less a public enlightenment or mass mobilization model of public relations, according to Nwosu (1996).

In the RICEE model, each of the components plays an indispensable role towards achieving the much-required result of healthy environment.

Research, as the name implies, would aid in gathering the relevant facts and figures as regards the kind and nature of environmental pollution in Enugu metropolis. Equally, data as regards the root causes of environmental pollution could be elicited through research. Such data as those related to:

- Rapid and wasteful use of resources with emphasis on pollution prevention and waste reduction.

- Failure of economic and political system to encourage earth sustaining forms of economic development and discourage earth-degrading forms of economic growth.

- Failures of economic and political systems to have market prices, including the overall environmental cost of an economic good or services.

- Our urge to dominate and manage nature for our use with far too little knowledge about how nature works.

- Simplification and degradation of parts of the earth's life support system (Miller, 1998:16).

More still, data as regards the indicators on environmental pollution such as:
- Neatness of the environment;
- Description of life-support systems for humans and other species;
- Damage to property;
- Nuisance such as noise and unpleasant smells, taste and sight can only be ascertained through research, which is an indispensable component of the PR RICEE model.

Furthermore, research will help us to elicit accurate information as regards the factors which determine the severity of the harmful effects of the pollutants. Such factors include:

- The chemical nature of the pollutant;

- Its concentration;

- And the pollutants persistence (Miller, 198:10)

On another note, the Information component will ensure that there is a reliable system for disseminating regular, adequate and appropriate information to the relevant publics as regards current ways of preventing or controlling environmental pollution.

More still, the Communication component, which as it were, is the centre of the enlightenment, guarantees the two-ways communication needed for effective management of the environmental degradation. According to Canfield and Moore, communication is a special process which makes human interaction possible and enables man to be a social being without which misunderstanding and conflict may result (Canfield and Moore, 1977:50). Communication, therefore, is a means by which one person influences, persuades or convinces another, an important tool with which he organizes, stabilizes and modifies his life as a member of a larger society (Siebert, 1977:15).

Although there may be differences in applications of communication, the process of communication between individuals and groups still remains basically the same. In its simplest form, the communication process involves the sending of a message through a chosen channel with the message couched or as the message is interpreted by the receiver. Communication can then be said to have taken place, though feedback may or may not have been sent. However, with regards to the PR RICEE model, feedback is greatly needed to really fine-tune the strategies towards achieving the much-needed healthy environment.

Yet on another note, the Education component ensures indepth and systematic education and public enlightenment programme, workshops and seminars aimed at creating the much-needed knowledge as regards the plagues and ugly consequences of environmental degradation. This will serve not only as deterrent to further pollution of the environment but would also help to inculcate overall healthy environmental habits on the people. The education component will equally involve the production and use of various teaching aids and methods that will promote acquisition of new knowledge, developing of new attitudes and possible changes of behaviours in relation to environmental issues and problems.

Thus, as could be vividly seen from the above explication, all the components of the PR RICEE model need to complement each other towards realizing the enlightenment campaign objectives.

Hence, the thrust of this case study is to evaluate the PR RICEE model to ascertain how best it could be applied as an enlightenment strategy towards tackling the problem of environmental degradation in Enugu metropolis.

Statement of the Problem

It is evident that the government of Enugu State, through the agency of the Enugu State Environmental Protection Agency, has been working hard to reduce environmental degradation in the metropolis of Enugu. Equally, other Non-governmental agencies are working towards curbing the plagues of pollution. But we are yet to experience a radical turn-around as regards achieving a highly pollution free environment.

It is against this background that the researchers wonder what options or alternatives could be used or applied to decisively combat this looming evil of environmental pollution in the metropolis of Enugu. Hence from the foregoing, it is appropriate to note that without proper enlightenment campaign against environmental pollution in Enugu metropolis, there is doubt as to whether there would be positive change of behavior by the inhabitants towards achieving a healthy environment. Thus the problem arises: what role could the Public Relations RICEE model play as an enlightenment strategy towards combating the problems of environment degradation in Enugu metropolis? If there is any role, how could it be applied and to what extent could it serve as a sustainable solution towards ensuring reduction of ignorance and promotion of environment friendly behaviours and attitudes.

Objectives of the Study

This study has been designed to achieve the following objectives:

i.   To verify the possibility of applying the PR RICEE model as a public enlightenment strategy for reducing environmental degradation in Enugu Metropolis.

ii. To identify and assess the effectiveness of those components of the PR RICEE model that makes it relevant in combating the menace of environmental pollution in Enugu metropolis.

iii. To determine how these components of the PR RICEE model can best serve as tools for enlightenment campaign towards reducing environmental degradation in Enugu metropolis.

iv. To identify ways through which the effectiveness of this PR RICEE model as a public enlightenment strategy could be measured.

v.   To proffer through the instrumentality of the PR RICEE model, a comprehensive enlightenment strategy/design or package for tackling and winning the campaign against environmental degradation in Enugu metropolis.

Enugu is the capital of Enugu State in South-Eastern Nigeria. it was the capital of the then Eastern Nigeria, the then East Central state of Nigeria, the then Anambra State of Nigeria, and the capital of the then Biafra. It is therefore, a very important urban town or city in Nigeria.

190

Research Questions

These questions, which form the crux of our study, would aid in carrying out a comprehensive survey:

i.   Can the PR RICEE model be effectively employed as public enlightenment strategy for reducing environmental degradation in Enugu metropolis?

ii. Can the PR RICEE model serve as an operational model for the application of public relations techniques and strategies to the management and promotion of environmental issues?

iii. In what ways could the effectiveness of PR RICEE, model, as an environmental public enlightenment strategy, be measured?

iv. Can the PR RICEE model really help us to come up with a more comprehensive public strategy/design or package for winning the campaign against environmental degradation in Enugu metropolis?

v.   Must all the components of the PR RICEE model be used together as a package in order to win the enlightenment campaign against environmental pollution?

HYPOTHESES

The following null hypotheses have been formulated and will be tested in this study:

HO_1 The PR RICEE model is not an effective public enlightenment strategy for executing public enlightenment campaign against environmental degradation in Enugu metropolis

HO_2: Research and Information which are components of PR RICEE model are not relevant enlightenment tools for addressing the problem of environmental pollution in Enugu metropolis.

HO₃: Two way integrated communication as a constituent of PR RICEE model cannot contribute to enlightenment campaign aimed at reducing environmental degradation in Enugu metropolis.

HO₄: PR RICEE component of educational execution cannot make any significant contribution in the public enlightenment campaigns aimed at reducing environmental degradation in Enugu metropolis.

Significance of the Study

The issue of environmental enlightenment, management and control towards ensuring a healthy environment is one which must be taken seriously since degradation of our environment tantamount to the destruction of the very thing upon which human existence depends. This study therefore, seeks to test of the PR RICEE model, to find out how applicable or useful it is in addressing environmental pollution problems using environmental degradation in Enugu State as a case study. It will no doubt contribute positively to environmental management policies, projects and research.

Scope of the Study

The public relations RICEE model would be analyzed or examined in this study as a public enlightenment strategy that can be used to address both short-term and long-term promotion of campaigns against environmental degradation problems in Enugu metropolis.

The research component of this Public Relations RICEE model will be tested to find out if and to what extent it can help us to gather and analyze relevant facts and figures that will help anyone in carrying out any public enlightenment campaign on the environment.

Equally, the information component will be tested to find out if and to what extent it can serve as a reliable system for disseminating regular, adequate and appropriate information to the relevant public.

More still, the communication ingredient of the RICEE model will be tested to find out if and to what extent two-way communication is needed to effectively carry out public enlightenment campaigns aimed at reducing environmental degradation in Enugu metropolis.

Also, the execution/education and evaluation components of this public relations model will be tested to find out if and to what extent they can back up the information and communication activities with in depth and systematic education, execution and evaluation strategies in environmental public enlightenment programmes.

Finally, adequate and comprehensive marriage or integration of all these components of the Public Relations, RICEE model will be analyzed to find out and to what extent it can be used to inculcate overall healthy environmental habits on the inhabitants of Enugu metropolis.

Research Methodology

This study adopted the survey research method and design for, according to Ozongwu Maurice Odo (1992:42), "Survey research design deals with the practical application of the already standardized theories available in social and behavioural sciences". In addition, survey research "focuses on people, the vital facts of people and their beliefs, opinions, attitudes, motivation and behaviours" (Osuala, 1991:181). Therefore, this form of research is appropriate in this study.

Sources of Data

Two main sources of data were used for this study.

Primary Sources

These are data sources originally and specifically designed and collected for this study by the researchers and which have not been used elsewhere by other researchers.This primary sources include questionnaire and personal oral interview. The questionnaire was designed to elicit specific responses from public relations mangers and officers with regards to answering the research questions and testing the hypotheses formulated in this study. Personal interviews were conducted by the researchers among some of the respondents to complement the data available from the questionnaire.

Secondary Sources

These were sources other than primary sources, which the researchers relied upon for additional materials to support the information available to them from the primary sources of data. These secondary sources included newspapers, magazine, newsletters, books and journals on public relations and also books and journals on the environment, as well as other related areas of disciplines, government documents and regulations (see Appendices) were also utilized.

Study Population

The universe of interest here ideally would include all the people in Enugu or experts who are knowledgeable about public enlightenment campaigns, especially in the area of application of PR strategies or models in public enlightenment campaigns on the environment. But for reasons of geographical, financial and time constraint, the study focused on public relations, environmental and other experts in three zones of Enugu metropolis viz; Enugu North, Enugu South and Enugu East, who know or should have adequate knowledge about public relations and the environment. The population size was not definite.

Sample Size Determination

Having defined the population, the researchers then proceeded to determine the size of the sample. Since no researcher really needs to interview and administer questionnaire to all the units in a very large population, the concept of sample size is used to overcome the problem.

Thus, the views and opinions of the public relations environmental officers and other managers in the three zone of Enugu metropolis as mentioned earlier, were examined in a pilot study. Factors such as the nature of the population, the degree of precision in sample size selection and other considerations by the researchers guided us in selecting the sample size. Based on all these, the following statistical formula was used in our effort to determine the final sample size for this study:

$$n = \frac{Z^2 \, (pq)}{e^2}$$

Where:

n = Sample size

z = Desired level of confidence

p = The probability function of the estimated Standard deviation

(i.e. Proportion of success)

p = P (Proportion of failure)

e = Estimated Standard Error

Working on a 95% confidence level and an estimated standard deviation of 80% with a tolerable error of 5% we got the following sample size:

z = 95% or 1.96

p = 80% or .80

q = 20% or .20

$\therefore$ n = $\dfrac{(1.96)2(80.20)}{5^2}$

= $\dfrac{6146.56}{25}$ = 246

Table 10.1:   Questionnaire Distribution in Aggregation

Geographical Location	Number of Questionnaire
Enugu North	83
Enugu South	93
Enugu East	68
Total	246

Therefore 246 respondents were administered with questionnaire as shown in Table 10.1. The questionnaire administration copies and oral interviews were conducted or carried out by the researchers.

Method of Data Presentation and Analysis

Simple frequency tables were used to present results of the questions in the administered questionnaire. Equally, simple percentage ratios were applied in presenting the results of data collected:

F/N x   100/1

Where F =   Frequency

N =   The summation of the frequencies

Also, the Chi-square (X^2) statistics was employed in the test of the hypotheses

The Chi-square formula is as follows

X^2 = (Oi-Ei)

(Ei)

Where: O =   Observed frequency

E =   Expected value of frequency

Data Presentation and Analysis

The data collected from respondents to this study are presented and analyzed in this section. One 31-items questionnaire was administered on the respondents after the questionnaire was tested for validity and reliability. 200 out of the 246 questionnaire copies administered by us were found to be usable, giving us adequate or reasonably high returns enough for the purpose.

Table 10.2:   **Recovery Rate of Questionnaire**

Geographical Location	Number of Questionnaire Administered	No. of Questionnaire Returned and Accepted	No. of Questionnaire Rejected	% Responses Rate
Enugu North	85	75	10	88%
Enugu South	93	74	19	80%
Enugu East	68	51	17	75%
Total	246	200	46	81%

As could be seen clearly from Table 10.2, a total of 246 questionnaire copies were administered, 200 representing 81% of the sample size were recovered in good condition, while 46 representing 19% of the simple size were either wrongly completed and therefore were rejected or not returned at all. Thus, this shows a high or acceptable response rate.

Table 10.3:   **Determination of Educational Attainment of Respondents**

Educational Attainment	Frequency	Percentage
Secondary	50	25%
Tertiary	95	47.5%
Postgraduates	55	27.5%
Total	200	100%

From the data available in Table 10.3, 50 or 25% of the sample size claimed to have attained secondary education while 95 or 47.5% claimed to have attained tertiary education. More still, 55 respondents representing 27.5% of the sample size claimed to have attained postgraduate education.

The predominance or higher percentage of tertiary and postgraduate respondents could be understood and explained from the standpoint that the respondents whose views were sought were of professionals not quacks.

Data Presentation Based on the Research Questions

In this section, we shall present the results or data that offer answers to the research questions we posed at the beginning of this study: -

Research Question Number One

Can the RICEE model be gainfully employed as an effective public enlightenment strategy for reducing environmental degradation?

Table 10.4:   Determination of the Suitability of the RICEE Model as an Effective Public Enlightenment Strategy For Reducing Environmental Degradation:

Geographical Location	YES	% Yes	N	% No	Neutral	% Neutral	Total	% Total
Enugu North	45	60|%	8	10.7%	22	29.3%	75	100%
Enugu East	50	67.6%	15	20.3%	6	12.1%	74	100%
Enugu South	33	64.7%	7	13.7%	10	19.6%	51	100%
Total	128	64.2%	30	15%	41	2.3%	200	100%

Table 10.4 which gives the picture of the respondent's answers at a glance, shows that of the 75 respondents questioned from Enugu North, 45 people representing 60 percent of the respondents from that zone gave positive answer, 8 people gave outright negative answer while 22 people representing 29.3% of the sample size from that zone remained neutral.

Equally, 50 people representing 67.6% of the sample size from Enugu East gave "yes" as their answer, 20.3% of the people numbering 15 respondents gave no an answer while 9 people representing 12.15 of the sample size from the sample size from zone were neutral to the question.

Furthermore, of the sample size from Enugu South Zone, 7 respondents gave negative response, 10 people representing 19.6% of the sample size took a neutral position, while 33 respondents representing 64.7% of the respondents from that zone confirmed the suitability of the RICEE model by giving positive responses.

Hence, of the 200 respondents in the three zones who responded to the above questions, 128 respondents representing 64.2 of the entire sample size gave positive response by answering "Yes" while 30 people representing 15% of the respondents responded "No" with 41 people representing 20.3% of the sample size remaining neutral.

Research Question Number Two

Can the PR RICEE model serve as an Operational model for the application of PR Techniques and Strategies to the Management and Promotion of Environmental Issues?

Table 10.5: **Determination of Whether the PR RICEE Model Can Serve as an Operational Model for the Application of PR Techniques and Strategies to the Management and Promotion of Environmental Issues:**

Geographical Location	YES	% Yes	N	% No	Neutral	% Neutral	Total	% Total
Enugu North	49	65.3%	20	26.7%	6	8%	75	100%
Enugu East	50	67.6%	11	14.8%	13	17.6%	74	100%
Enugu South	33	64.7%	10	19.6%	8	5.7%	51	100%
Total	132	66%	41	20.5%	27	13.5%	200	100%

From Table 10.5, it is evident that 49 respondents representing 65.3% of the sample size from Enugu North agreed that PR RICEE model can serve as an operational model for the application of PR RICEE techniques and strategies to the management and promotion of environmental issues and problems. Also 20 people representing 26.7% gave negative answer with 6 respondents representing 8% of the sample size from the zone remaining neutral.

On the other hand, 50, 11 and 13 respondents representing 67.6%, 14.8% and 17.6% of the sample size from Enugu East responded Yes, No and Neutral respectively.

Thus, 132 people representing 66% of the entire sample size affirmed the capability of the PR RICEE Model to serve as an operation model for the application of PR techniques and strategies to the management and promotion of environmental issues and problems, while 41 people representing 20.5% of same population answered No to the question with 13.5% numbering 27 respondents remained neutral.

Research Question Number Three

In what ways could the effectiveness of the PR RICEE Model as an environmental public enlightenment strategy be measured?

Table 10.6: Determination of the Way Through which The Effectiveness of the PR RICEE Model as an Enlightenment Strategy could be Measured

	Neatness of the Environment				Environmentally Sustainable Attitude/Behaviour				High level of Awareness among the Populace of Environmental Issues					Execution of Environmental Projects					Geographical Location
	Yes	% Yes	No	Neutral	Yes	% Yes	No	Neutral	Yes	% Yes	No	% No	Neutral	% Neutral	Yes	% Yes	No	Neutral	
Enugu North	75	37.7%	-	-	75	37.7	-	-	45	30.3%	5	13.9%	25	62.5	75	37.5%	-	-	Enugu North
Enugu East	74	37%	-	-	74	37.00	-	-	39	31.4%	20	55.6%	15	37.5	74	37%	-	-	Enugu East
Enugu South	51	25.5%	-	-	51	25.50	-	-	40	32.3%	11	30.5%	-	-	51	25.5%	-	-	Enugu South
Total	200	100%	-	-	200	100%	-	-	124	100%	36	100%	40	100%	200	100%	-	-	Total

The data collected as regards the research question 3, shows that 100% which implies all the respondents numbering 200 and making up the entire sample population, indicated that neatness of the environment, environmentally sustainable attitude, behaviours and execution of environmental codes/ethics and laws are all veritable measurements for the effectiveness of PR RICEE Model.

On another note, 45, 39, and 40 respondents representing 30.3%, 31.4% and 32.3% from Enugu North, East and South respectively maintained that high level of environmental awareness of the PR RICEE Model as an enlightenment strategy. But 13.9%, 55.6% and 30.5% numbering 5, 50 and 11 respondents from Enugu North, and Enugu East and South respectively answered No to the same option.

However, 25, and 15 people representing 62.55% and 37.5% from Enugu North and East respectively remained neutral to this option of high level of environmental awareness among the populace.

Research Question Four

Can the PR RICEE Model really help us to come up with a more comprehensive public enlightenment strategy /design or package for winning the campaign against environmental degradation in Enugu metropolis?

Table 10.7: **Whether the PR RICEE Model can really help us to get a comprehensive Public enlightenment strategy/design or package for winning Environmental Degradation Campaign**

Geographical Location	Yes	% Yes	N	% No	Neutral	% Neutral	Total	% Total
Enugu North	49	65.3%	10	13.3%	16	31.4%	75	100%
Enugu East	50	67.6%	12	16.2%	12	16.2%	74	100%
Enugu South	35	38.6%	6	11.8%	10	19.6%	51	100%
Total	134	67%	28	14%	38	19%	200	100%

202

As can be seen from table 10.7 above, of the 75 respondents questioned in Enugu North, 49 people representing 65.3% of the sample size from that zone gave positive response, 10 persons representing 13.35 respondents answered negatively, while 16 people representing 31.4% of the sample size from the zone remained neutral. Equally 50, persons representing 67.6%, 16.2% responded Yes, while 12 persons representing 16.2% responded No, and another 12 persons representing 16.2% remained neutral.

Morestill, of the 51 people administered questionnaire in Enugu South, 68.6% numbering 35 persons responded Yes, 11.8% numbering 6 persons gave NO as answer while 10 people representing 19.6% of the entire sample size from that zone were neutral. Hence from the entire data, we see that 134 respondents representing 67% of the entire sample size gave positive response.

While 28 people representing 14% of the sample size stated No as their answer, with 19% numbering 38 persons remaining neutral.

Research Question Number Five

Must all the components of the PR RICEE Model be used together as a package in order to win the enlightenment campaign against environmental pollution?

Table 10.7:   **Whether the PR RICEE Model can really help us to get a comprehensive Public enlightenment strategy/design or package for winning Environmental Degradation Campaign**

Geographical Location	Yes	% Yes	N	% No	Neutral	% Neutral	Total	% Total
Enugu North	70	93.3%		-	5	6.7%	75	100%
Enugu East	62	83.3%	3	4.1%	9	20.3%	74	100%
Enugu South	39	76.5%	5	9.8%	7	3.7%	51	100%
Total	171	85.5%	8	4%	21	10.5%	200	100%

As could be vividly seen from Table 10.8, 70 respondents representing 93.3% of the sample size from Enugu North were of the opinion that all the components of the PR RICEE Model be used together as a package in order to achieve the needed enlightenment campaign objectives. No respondent objected to the contrary. However, 6.7% of same sample size numbering 5 persons remained neutral to the question.

Furthermore, 83.8% of the sample size from Enugu East numbering 62 respondents agreed to the view that the PR RICEE Model must be used as a complete package while 3 people representing 4.1% of same sample size answer No. with 9 people representing 20.3% of the sample size remaining neutral.

Also, 39.5% and 7 respondents representing 76.5%, 9.8% and 13.7% respectively responded Yes, No and neutral in Enugu South. Thus, when we consider the entire sample size, they affirmed the view that all the components of the PR RICEE Model must be used together as a complete package in order to win the public enlightenment campaign against environmental degradation/publication. On the contrary, a total of 8 people representing 4% of same sample size numbering 21 respondents maintained a neutral position.

Hypotheses Test Results

Five hypotheses were formulated and tested in this study using the chi-square test statistics. The findings or results from the test are reported below.

Hypothesis

Our test of this hypothesis revealed that the X^2 calculated was greater than X2 Tab. We therefore rejected the Null Hypothesis (Ho) we stated earlier in this chapter and accepted its Alternative (Hi) which means that the PR RICEE Model is an effective public enlightenment strategy for executing campaigns against environmental degradation. (X^2 cal. Was 15.43 while the X tab was 9.49).

Hypothesis 2

Our test of the second hypothesis also showed that the X^2 cal was greater than the X^2 tab. So, we rejected the Ho or Null hypothesis we stated earlier in this chapter and accepted its alternative, which led us to conclude that in PR public enlightenment campaigns, the Research and Information components of the PR RICEE Model are relevant public enlightenment tools for addressing the problem of environmental degradation in Enugu metropolis. (X^2cal. = 10.3 X^2 tab = 9.49).

Hypothesis 3

Again, from our test of hypothesis Number 3 which we stated in the Null form earlier in this chapter, we found out the X^2 cal. was greater than the X^2 tab. (i.e. 19.7>9.47). we therefore rejected the Null hypothesis and accepted its alternative, which led us to the conclusion that two-way integrated communication, which is a component of the PR RICEE Model, contributes significantly to public enlightenment campaigns aimed at reducing environmental degradation in Enugu metropolis.

Hypothesis 4

Similarly, in our test of the Fourth Hypothesis which we stated in the null form (Ho) earlier in the chapter, we discovered that X^2 cal. was again greater than X^2 tab. (i.e. 15.2>9.49). So, we concluded that the PR RICEE components of execution/education and evaluation can make significant contributions to public enlightenment in Enugu metropolis. In other words, we rejected the Null hypothesis and accepted its alternative.

Discussion and Conclusion

This section discusses the results of the research effort. Equally recommendations were made and the study is ended with a conclusion.

Summary of the Findings

The following major findings were made:

(a)   Ignorance of the dangers and evil consequences of environmental degradation is itself a major cause of the persistence of the phenomenon.

(b)   The PR RICEE model is an effective public enlightenment strategy for executing the campaign against environmental degradation.

(c)   Research and information which are the components of PR RICEE model are very relevant PR public enlightenment tool for addressing the problems of environmental pollution.

(d)   Two ways integrated communication, which is also a component of the RICEE model, is simply indispensable in public enlightenment campaigns aimed at reducing environmental degradation.

(e)   PR RICEE component of execution/education and evaluation also make significant contribution in designing and carrying out EPR public enlightenment campaign.

(f)   The respondents believe that effectiveness of the PR RICEE model as a Public Relations enlightenment Strategy, could be measured through viz:

    i. General neatness of the environment

    ii. Environmentally sustainable attitudes/behaviours

    iii. High level of environmental awareness by the population

    iv. Diligent execution or implementation of environmental codes/ethics and laws.

(g)   Most respondents sampled maintained that the mass media including radio, TV, newspapers and magazines, are the most effective way to reach out to the public in the awareness level of the campaign.

(h)   As at present, the level of environmental degradation in Enugu metropolis is still relatively high.

Discussion of the Findings

The fact that people do not normally act on what they are not aware of, unless perhaps by some stroke of chance, goes to buttress the findings that ignorance as regards the evils of environmental degradation is itself a major cause of its persistence. Proper and convincing knowledge is what usually leads people to action. Thus, the researchers vehemently uphold the fact that if the issues and problems of environmental degradation in all their ramifications are correctly and comprehensively related to the populace, it would facilitate prompt and encouraging elicitation of co-operation from the public towards solving the problem.

Hence, this brings us to the crux of our finding, which is, that PR RICEE model is an effective public enlightenment strategy for executing the enlightenment campaign against environmental degradation. The PR RICEE model is a proven activity-oriented public relations strategy, and according to David Edeani, public relations activities do influence the formation of public opinion. He observes, that the relationship between public relations and public opinion on the one hand and public opinion and attitude changes on the other, is overwhelmingly reciprocal in character (Edeani, 1993:109). Thus, public relations could be viewed as a vehicle for communicating and changing public attitude. Therefore, through proper application of the PR RICEE model as a PR public enlightenment strategy, the people will be adequately conscientized, informed and educated. This would abundantly lead to a change of attitudes, opinions and behavior. As researchers have abundantly demonstrated, everything being equal, a change in opinion will involve a corresponding change in attitude and behavior.

Furthermore, the findings on C and E above, go to make clear the view that all the components of the PR RICEE model are very relevant for a comprehensive execution of the much needed public enlightenment campaign aimed at reducing environment degradation. Thus, the RICEE model must be understood and applied as a complete package, for it to be really efficient and effective. This is because each of the components, Research, Information, Communication, Education/Execution and Evaluation, will complement each other, towards the realization of the objective of any EPR enlightenment campaign.

The finding that the respondents believe that the effectiveness of the PR RICEE model as an EPR enlightenment strategy could be measured by neatness of the environment, diligent execution of environmental codes, laws, and environmental sustainable attitudes, goes to strengthen the fact that public relations enlightenment campaigns are aimed at achieving concerted practical results. The enlightenments are not done for enlightenment sake; they are directed towards pragmatic objectives. Hence, this corroborates Philip Lesly's view that besides publicity, public relations people determine what they must do to get the goodwill of others, it plans ways and means of winning that goodwill and it carries out activities designed to win it, and consequently achieve the results needed (1998:6).

Still discussing the measurement of the effectiveness of PR RICEE model, we noticed in the course of the research that many respondents ticked high level of awareness together with environmentally sustainable attitudes/behaviour. This further buttressed our earlier assertion that there is a link between education, awareness and action, which in this case could be termed change of behavior. Hence, when the populace have been adequately enlightened, conscientized and persuaded through well-articulated and comprehensive EPR enlightenment and other EPR activities, goodwill is consequently elicited from the people in the form of change of attitudes and behavior favourable to or in favour of the enlightenment objectives.

On another note, when we consider the endorsement of the use of the mass media, which includes radio, TV, newspapers, magazines and advertising by most respondents, as the most effective way to reach the public in the course of the enlightenment campaigns, we find out that their stand is typical of a country like ours which is still developing or which some people would call "third world countries". This stand, which they maintained regarding the mass media is not unconnected with the fact that majority of our population are not educated. Hence they find it too academic to attend public lectures, workshops and seminars. Morestill, publications on the state of the environment would only be accessible to the literate people who equally have the fund to subscribe. And most of our people have not come to that level of enlightenment to consider such publications relevant.

Thus, when we consider the characteristics of the audience constituting our population, the channel or media of RADIO and perhaps televisions, seems even the most appropriate and most effective medium of communication that plays a significant role in ensuring that messages are adequately communicated. In fact, Marshall Macluhan, the late Canadian communication expert, considers the medium so central to effective communication, that he made the now classical but controversial statement that "the medium is the message". This is said to be his way of perhaps pointing out that the medium employed in any communication situation usually has positive or negative effect of influence on the other key elements in the communication process; the audience and the communicated message itself. We must not forget, however that in any EPR campaign, we must include the traditional media of communication, (the oramedia) in our mix for any rural EPR campaign that is aimed at or includes the rural areas where more than 70% of Nigerians or Africans still live.

The finding that the level of environmental degradation in Enugu is still high, goes to buttress the stand of the researchers who equally share the same view. Furthermore, it equally gives meaning and relevance to this research initiative, which is seeking a PR option to address the problem. This is simple because, if there is no problem there would be perhaps no need for a solution.

Recommendations

Based on the findings, the following recommendations are hereby made:

1.   In order to solve the problem from the root, the issues of ignorance must be addressed properly. This could be achieved by ensuring that the evil consequences and dangers arising from environmental degradation are well brought into focus while articulating a comprehensive PR enlightenment campaign.

2.   The PR RICEE model, having been proven as a veritable PR public enlightenment strategy for addressing environmental problems and issues, is hereby recommended to all stakeholders in the management of our environment. They are however, to modify it contextually from circumstance to circumstance and from problem to problem, to achieve the much needed rewarding results.

3.   We equally strongly recommend that this PR RICEE model be adopted as a complete package, not leaving out any of its components in the enlightenment campaign so as to get the required effective results.

4.   Considering the characteristics of most audience constituting the population in our environment, we recommend that such media as radio, television and daily newspapers that circulate extensively within our urban cities be used mostly for reaching out to the public as these are media that are more accessible to the populace in this part of the world. And for those living in rural areas, we recommend the use of mainly the traditional, interpersonal and group media of communication.

5.   We equally recommend that the emphasis of the enlightenment campaigns should not be on threats of prosecution against defaulters of environmental regulations, but on conscientization aimed at convincing the people that to act in an environmental-friendly way is for their own good and that they need not to be policed to act accordingly. This is in line with public relations preference for persuasion as opposed to coercion. But sometimes a planning admixture of coercion and persuasion may work better in some situations.

6.   Considering the fact we discovered in our research that habits and things which people learn when they are younger easily stay with them through life, we strongly recommend that comprehensive environmental education should be made a compulsory part of early school curricular, and even up to the tertiary school level so that the culture of environmental-friendly behaviours would be inculcated into the children early enough.

7.   Efforts must be made by those charged with the management of our environment to ensure that facilities are put in place to help people to conveniently dispose of their refuses. Morestill, collected

refuses should be timely evacuated from living vicinities to maintain a healthy environment.

Conclusion

For a proper management of our environments towards maintaining a healthy and sustainable global environment, it is now evident that a well-articulated and comprehensive EPR Public enlightenment campaign is indispensable. This is because the support and co-operation needed from the populace in order to achieve this sustainable environment will usually be elicited only when the populace have been properly enlightened, educated and conscientized on the evil consequences of environmental degradation and that we cannot afford to exploit our environment without check. Hence, the public relations enlightenment strategy proposed in this research-based chapter to adequately achieve this enlightenment campaign is the PR RICEE model originally propounded by Prof. Ikechukwu E. Nwosu. And it is our candid submission that if all the findings and strategies articulated in this chapter are put into effective use, healthy environmental attitudes and behaviours would be inculcated into the general populace and this, in due time, will naturally translate into a healthy and sustainable environment.

Chapter Eleven

Media images of environmental issues and problems in Nigeria
- A Content Analytical Case Study -

Introduction

In recent times, the management of the volatile environmental issues through Public Relations has come to the center-stage of modern public relations practice. Public relations managers, consultants and researchers the world over are therefore currently engaged in the search for knowledge, skills, strategies, data and techniques for the effective application of PR to sustainable development.

The review of literature for this study showed that there are not yet much published works that are based on research on environmental public relations in Nigeria and even Africa. The International Public Relations Association (IPRA) had, in its 1992 *Gold Paper No. 9* on "Green Communications and sustainable development" by Bruce E. Harrison (1993), offered an indepth analysis of what should be the role of public relations managers in environmental protection and sustainable development. Among other things, the Gold Paper stressed the need for continued research on various aspects of this area of professionalism in public relations. It also pointed out the need for public relations managers to be at the same time *proactive, reactive* and *interactive* in managing environmental issues and problems.

This IPRA Gold Paper also offered insights on what should be the role of all communicators (including the mass media operators) in handling the important issues and problems involved in environmental protection and sustainable development.

Another work by Anne Onumonu(1996:32-37) which was done in the Nigeria and African context endorsed most of the points made above by the IPRA *Gold Paper* No. 9, and was published by the federation of African Public Relations Association (FAPRA). It reviewed extensively the contemporary environmental issues and problems in Africa and recommended, among other things that corporate bodies operating in Nigeria and other African countries "must come to realize that they have to face trade-offs" when dealing with environmental issues and problems in their areas of operation, instead of concentrating only on conventional economic activities. Among the key environmental issues in contemporary Africa indentified by the researcher, were land use/degradation, pollution, toxic wastes disposal, population, poverty, hunger and debt burden, technology transfer and cooperation. It will be interesting to find out in the present study what the Nigerian Mass Media studied consider to be the key environmental issues and whether they will be similar to the above-listed issues.

One other insightful work on public relations and the environment that was done in the African context is the one by Ugandan expert F.R. Turyatanga (1996:23-26). The issues and problems on the environment which he said public relations managers must help to tackle were deforestation, soil erosion, over – grazing, poverty, drought and desertification, malnutrition, infant mortality, dumping of toxic waste, pollution of air, soil and water, the arms race especially arms of mass destruction, loss of biological diversity, depletion of the ozone layer, global warming and climatic changes, green house gases, economic pressures and the consequent overexploitation of the environment. He recommended that public relations managers in Africa must not only endeavour to be knowledgeable on these environmental issues and problems, but should be able to educate others in and outside their organizations on these issues and problems. They must also strive to prevail on their institutions, identify with environmental issues and problems, make their corporate policies sensitive to these problems. The role of the mass media in helping the public relations manager to do these, though not directly mentioned by Turyatanga, can be deduced from his work. The present study is on this subject matter and more.

It must be noted, however, that none of the works reviewed above so far was a quantitative analysis of the environmental issues and problems. Some quantification, as in the present study, will help Public Relations managers to be more precise, authoritative and confident in handling environmental issues and problems for their various organizations in the continent.

Conspicuously absent in the literature, are studies on media images of environmental issues from the public relations perspectives in African or Nigerian contexts. There are however, a number of large unpublished studies on this subject in the literature on the general question or area of mass media coverage of the environment, done purely from journalistic perspective. Among such studies are the ones by Olugbemi (1990) and Akpa (1995). Both studies used content analysis to systematically study selected Nigerian newspapers' performances in their coverage of environmental issues and problems. Both studies discovered that the newspapers they studied gave inadequate coverage to environmental issues and problems in Nigeria. Akpa's key conclusion went thus:

> "These findings reveal that newspapers coverage of events and issues in Nigeria has not specialized to the extent of giving adequate coverage to environmental issues and problems in Nigeria"

It will be interesting to compare this verdict on the same matter, even though we were more concerned here with the implications of such verdicts for public relations practice.

215

On general issues of environmental protection and sustainable development, our review of the literature shows that since they came in 1972 during the first United Nations Conference on Human Environment, environmental protection and sustainable development has received quite a lot of research and professional attention, with many published and unpublished works emerging on those issues from various fora and regions in the world. This goes to show the critical and topical nature and importance of the subject matter under study in the present research effort.

The issues and problems of man's environment was considered so important by the U.N. General Assembly that in 1984, it set up a body known as the world Commission on Environmenta and Development (WCED) to formulate a global agenda for positive changes in this area. Back home in Africa, the first African Ministerial Conference on the Environmental (AMCEN) held in Cairo, Egypt, came up with what they called the Cairo Programme for African Cooperation, which among other things emphasized the need for backing environmental degradation, enhancing food production, rectifying the imbalance between population and resources, and achieving self-sufficiency in energy in African countries (Onumonu, 1996).

However, the literature research has revealed that the one single event that brought men in all societies to the realities of not protecting the environment or its consequences, is the Earth Summit held in Rio de Janairo, Brazil in 1989 and attended by 182 countries. This was the second United Nations Conference on the Environment and Development. The conference came up with a Plan of Action, commonly known as Agenda 21, for the effective management of the pressing environmental problems of today and preparing us even for the challenges of environmental protection and sustainable development in the new millennium (the 21st century). It is significant to note that the first African workshop on the implementation of Agenda 21 was held in Abuja, Nigeria in January 1993 (Onumonu, 1996). Yet, as shown in the literature, the Nigerian people and the Nigerian Mass

Media do not seem to be as sensitive as they should be to contemporary environmental issues and problems in the world today. This has serious implications in this study, which touches on the mass media role on this subject matter and focuses on the implications of this role for Public Relations Practice and Public Relations managers.

Problems Statement

Public relations managers do not yet seem to be quite sure of the nature or pattern of coverage given to environmental issues and problems by the mass media in Nigeria or the images of these problems portrayed by the mass media. They cannot afford to depend on guesswork or hunches in this sensitive area of environmental public relations. They need concrete data to fill this dangerous knowledge gap in public relations professional practice in this country.

This study is an attempt towards helping to fill this knowledge gap. It is specifically an attempt to systematically describe and analyze the nature or pattern of coverage or reportage given to environmental issues and to bring out the implications of this pattern of coverage for public relations managers or practitioners in the country. This explanatory and descriptive study will intentionally focus first on the print media (newspapers) to generate baseline data that can be used to do more indepth and more extensive coverage of environmental issues and problems in Nigeria from public relations management perspective. The newspapers selected for this study are *the Daily Times and New Nigerian* representing the government-owned newspapers, and *the Guardian and Champion* representing privately owned newspapers. The study will cover a 12-month Audit period (August 1, 1996 – July 31, 1997).

Objectives of the Study

In addition to the above-stated general objectives, the following specific objectives have been stated to give further focus and "bite" to the study:

1. To find out what pattern of coverage is given to environmental issues and problems by the selected Nigerian newspapers.

2. To describe and analyze various dimensions of this pattern of coverage (e.g. quantity and quality of coverage and ownership influences, if any).

3. To examine and point out the implications of this pattern of coverage for the public relations manager or public relations practice in Nigeria.

4. To make appropriate practical public relations recommendations based on the above findings and analyses.

Research Questions

To give further focus to this exploratory study, the following research questions were posed and answers to them are expected to be offered by this study:

1. What are the frequency scores of the selected Nigerian problems within the 12-month period under study?

2. To what extent do the ownership patterns of these newspapers influence their pattern of coverage of environmental issues and problems (if it does)?

3. What is the collective placement or prominence score of these newspapers in their coverage of environmental issues and problems?

4. What is the collective interpretation score of the selected newspapers under study in their coverage of environmental issues and problems?

5. What are the practical implications of our findings on 1-4 above for public relations managers and public relations practice in Nigeria?

6. What practical public relations recommendations can be made, based on all these?

Scope of Study

Since this study is aimed at essentially generating baseline data as an exploratory study, we did not consider it necessary yet, to formulate and test any hypothesis in the study. The study is also designed to have more practical than theoretical application, even though we know that theory and practice are two sides of the same coin.

As noted above, four (4 leading) Nigerian newspapers (*New Nigerian, The Guardian, Daily Times* and *Champion)* have been selected for this study, which is intentionally limited for now to the print media. *The Guardian* and *Champion* are privately owned, while *New Nigerian* and *Daily Times* are government-owned.

The study covered a 12-month period for greater focus and control. Only newspaper contents that fit into our definition of environmental issues and problems were identified and analyzed. The implications of the findings of the study for public relations were also intentionally limited to areas of public relations management like community relations/corporate social responsibility, media relations, issues and crisis management and government relations which tend to touch more frequently on environmental protection and sustainable development issues and problems. The space measurement approach was not employed in this study because several past studies have shown that there is no significant difference in the findings when it is used along with the frequency or items count approach (Budd, 1967; Parker, 1979; Nwosu, 1995).

Limitations

The study faced the well-known constraints usually encountered in carrying out public relations and systematic social research (e.g. missing editions of some of the newspapers etc.) which are now very well known and documented (Sobowale, 1989; Nwosu, 1996; 78-79).

Methodology

The method of data collection in this study was systematic content analysis. The unit of analysis was the story. News and other items about environmental issues and problems within the study period (August 1, 1996 to July 31, 1997) in the 4 newspapers under study were read, categorized and coded for purposes of description and analysis in a manner similar to the ones used by Richard Budd (1964; 39) and Jim Hart (1961:54).

A total number of 1,544 editions of the newspapers constituted the population of the study. Out of this population, a total number of 120 editions were selected by using the Random Number Table and assigning 30 editions to each of the 4 newspapers studied. So, the quota and simple random sampling methods were applied in selecting the sample size of 120.

A code sheet and a set of coding instructions were developed in the data collection exercise. These were subjected to an Inter-coder Reliability Test. A relatively high inter-coder agreement of 88% between two independent coders (this researcher and a doctoral student of his) was achieved after a number of adjustments had been made on the code sheet and coding instructions.

Two sets of 15 items from the selected sample size of 120 above were used for the test. The inter-coder agreement was calculated using a combination of the simple percentage procedure and the following formula:

$$A = \frac{Po - Pe}{100 - Pe}$$

Where Po is the observed percentage of agreement, Pe is the expected agreement by chance, and A is the inter-coder agreement.

The key variables coded or measured in this study were frequency or number of items (to measure quantity), interpretation and placement (to measure quality of coverage) and the ownership categories. The sample size of 120 was considered adequate for this study whose population is 1,544, based on the widely accepted statistical rule that for a population of 1,000 to 2,000, 100 will be appropriate as sample size.

The data generated from the above coding exercise were computer-analyzed using the SPSS or Social Science package. The computer data collected were described, interpreted and further analyzed by the researcher. The findings or results from the entire exercise are presented below:

Findings/Results

Quantity of Coverage

This exploratory study came up with quite some interesting revelations on the pattern and nature of coverage given to environmental issues and problems in Nigeria. Firstly it made the startling discovery that in spite of the rapid growing rate of global concern for the environment, Nigerian newspapers are yet to start giving adequate coverage to these issues, both quantitatively and qualitatively.

Quantitatively, for instance, it is significant to note that for the 12-months covered by this study; the 120 editions of the 4 national newspapers studied carried only 74 items on the environment. Out of these 74 stories, *New Nigerian* reported 23 or 32%, *Guardian* followed with 20 stories or 28%. While *Champion* and *Daily Times* had 16 or 21% and 15 or 20% respectively. This distribution pattern is presented more on Table 11.1 below:

Table 11.1: Quantity of Stories on the Environmental Reported by Four Selected Nigerian Newspapers (% were rounded to remove rounding error)

S/No	Newspaper	Absolute Frequency	Relative Frequency
1.	*New Nigerian*	23	32%
2.	*Guardian*	20	28%
3.	*Champion*	16	21%
4	*Daily Times*	15	20%
	Total	74	100%

Government vs. Private Newspapers

On the question of whether government newspapers publish more or less stories on the environment due to the perceived leading role of government in carrying out campaigns on the environment, there does not seem to be any deducible significant difference in the quantity of coverages based on whether the newspaper is government-owned or privately owned. The data on Table 11.1 above confirms this. The slight edge, which the *New Nigerian* newspaper had over *Guardian* seems to be counter-balanced by the slight quantitative edge, which the *Champion* had over the *Daily Times*. We may therefore rightly say that the privately owned newspapers are as interested on the environment as the government newspapers. The verdict of quantitative coverage inadequacy for these two types of newspapers remains valid in this study.

Types of Environmental Issues Covered

Predictably, the newspapers studied published more stories on environmental sanitation and degradation than on any other issues or problem on the environment- A whopping 37 or 50%. Equally of the environmental stories published by the 4 newspapers, 16 or 22% were on Desertification, Soil Erosion- 10 or 13%, Deforestation- 6 or 8.0%. Note the complete absence of such global environmental issues as global warming due to the depletion of the ozone layers in the list of environmental issues focused upon by the newspapers examined and for the study period of one year. This is significant. Table 11.2 below presents the above data for easier perusal.

Table 11.2: Major Environmental Issues Covered By the Newspapers Analyzed

S/No	Environmental Issue	Absolute Frequency	Relative Frequency
1.	Environmental Sanitation & Degradation	37	50%
2.	Desertification	16	22%
3.	Soil Erosion	10	13%
4.	Deforestation	6	8%
5.	Pollution	5	7%
	Total	74	100%

(Data treated to remove rounding error)

The relatively high amount of coverage given to environmental sanitation and degradation by the 4 newspapers under study is no doubt as a result of the once a month environmental sanitation campaign carried out by the Federal Government and the twice-a-month environmental sanitation efforts carried out by some States and Local Governments in the country.

It is however, significant to note that this pattern of higher coverage for the environmental sanitation and degradation issues seems to be a general phenomenon world-wide. Joseph Kelly (1993:38) confirmed this in his study and pointed out that the mass media tend to give highest publicity to this issue "because of its centrality and importance in environmental reporting".

We should be happy that Nigerian mass media, based on the above observation, seem to recognize the most important environmental issue now (sanitation and degradation). But it is not a matter for cheers as the quantity of coverage given to it is still generally inadequate (37 stories by 4 papers in one year) as revealed in this study. We should also be disturbed by the fact that the other environmental issues like soil erosion, pollution, desertification and deforestation, which have continued to pose serious threat to man in Nigeria and Africa, received even much lower coverage as shown in Table 11.2.

And as noted above, the complete silence of the newspapers studied on the issue of global warming should be bothersome to us all because this is one of the key environmental issues in the world today. The implication is that they have not performed their social function of informing and educating their target audiences or publics on this critical environmental issues, just as they have grossly under-informed or under-educate them on the other environmental issues they gave low quantitative coverages to.

Quality of Environmental Issues Coverage

This exploratory study also revealed that the newspapers analyzed did not only give poor quantity coverage to the environmental issues they handled, but also gave them low quality coverage. In this study we used the variables of *interpretation and placement* to evaluate these newspapers' collective performance on the question of quality of environmental issues coverage.

On the question of placement or prominence, this study showed that most of the environmental stories covered by the newspapers were placed on their inside pages. Those stories were therefore, not given prominence because they were not strategically placed on the front or back pages of the newspapers studied. This is an indication of poor quality coverage. As revealed by the study, as many as 64 or 90% of the 74 environmental stories reported by these newspapers were placed on their inside pages. Only 4 or 4.5% of these stories made the front-page, while 6 or 5.5% of the rest of the stories were published on their back pages. The data reported on Table 11.3

in this study gives a more graphic picture of the above patterns in these newspapers' placement of stories on the environment.

If it is considered that the placement given to stories by the media shows the importance attached to them, the great import of the above data distribution is the poor performance of the 4 newspapers under study on the variable of placement.

Table 11.3: The Placement or Prominence Pattern of the Selected Newspapers Coverage of the Environment

S/No	Placement/Prominence Category	Absolute Frequency	Relative Frequency
1.	Inside pages	64	90%
2.	Front pages	4	4.5%
3.	Back pages	6	5.5%
	Total	74	100%

(Data treated for rounding errors)

So also the conclusion can be established that the importance, which our publics attach to these stories, is directly related to the importance or salience, which the media lend to these stories' according to the Agenda Setting Theory of the media (McComb & Shaw, 1973). It is also quite obvious that the placement of these environmental stories will determine their visibility and therefore determine whether they are read by the publics or not. All these explain why placement affects the quality of media coverage of events or issues.

On the question of interpretation and its influence on the quality of the environmental issues and problems covered by the newspapers under study, it was found out that the stories published by the newspapers were mostly straight interpreted news stories or "dead-pan" reports. For example, 31 or 44% of the stories reported were straight news stories while only 17 or 22% of the stories were feature stories, 16 or 21% were letters to the editor, while 10 or 13% were cartoons. There was not even a single Editorial on environmental issues and problems during the study period. There was also no supplement on the subject matter. These are why we passed a verdict of low interpretation and therefore low quality on the pattern of coverage given to environmental issues and problems by the newspapers under study. Most environmental issues and problems need to be properly interpreted by the media before they can be understood and appreciated by our publics or audiences, e.g. global warning and deforestation whose meanings and implications did not yet seem so clear to many Nigerians and other Africans. Un-interpreted stories also lack the hard-hitting or subtle persuasive contents, which will help to change the negative attitudes, opinions and behavior of many Nigerians and Africans to these critical environmental issues and problems of our time.

Table 11.4 below presents the collective interpretation scores or patterns of the four newspapers under study in their coverage of environmental issues and problems.

Table 11.4: The Performance of the Selected Newspapers on the Interpretation Variable

S/No	Interpretation variables	Absolute Frequency	Relative Frequency
1.	Straight News	31	44%
2.	Feature stories	17	22%
3.	Letter of the Editor	16	21%
4.	Cartoons	10	13%
5.	Editorials	-	-
6.	Supplements (Intra-media)	-	-
	Total	74	100%

(Data treated for rounding errors)

Discussion

The implications of the above findings on media images of or handling of environmental issues and problems for public relations managers and public relations practice in Nigeria are quite many. Some of them, in fact, can already be deduced. However, there is still the need for more pointed examination of some specific implications for public relations managers and public relations practice in this section. The areas of professional practice which this study has implications for, cut across the practical areas of community relations, corporate social responsibility, issues and crisis management, government relations, media relations and research in public relations. It even has implications for international public relations management even though this will not receive direct treatment or focus in this study.

Community Relations and Corporate Social Responsibility Management

A lot can be written on the implications of this study for community relations and corporate social responsibility management in Nigeria. This, as we know is a very sensitive and important area of public relations practice in which a lot has already been written.

This study has demonstrated that the community relations and CSR manager in Nigeria cannot depend entirely on mass media to educate themselves, and educate their target internal and external publics on contemporary environmental issues and problems. They need therefore, go beyond the Nigerian mass media to procure foreign publications in these subject areas, which as we have noted, are now many and varied, with the help of their organizations. They should also attend workshops, seminars and conferences on the environment to update their knowledge and skills on environmental protection and sustainable development management. This study has shown that the mass media in Nigeria not only under-cover environmental issues and problems, but have given low quality coverage to the little they publish. This says a lot about them and means a lot to the community relations and CSR manager in their attempt to keep improving their knowledge and skills on effective management of environmental issues and problems.

Media Relations Management

For the media relations managers, community relations managers and other managers who need to work effectively with the mass media to achieve their corporate objectives in the area of environmental protection and sustainable development, this study has a lot of implications. For one thing, they now should realize that in managing environmental issues and problems or communicating about them, they will be making a great mistake to use only the mass media (print or electronic). They need to use them along with other interpersonal, traditional and group communication networks, including computer-based ones (e.g. internet) where available. In doing this, they need to adopt a multi media approach always in managing their corporate communications on the environment.

Furthermore, they should use the so-called commercial news syndrome now in vogue in Nigeria (and even in Ghana) to get their news and other media materials onto the media, ensuring that they have got their management to make adequate budgetary provisions for this. In addition, these public relations managers should use more advertisements and advertorials to make up for the deficiencies of the Nigerian mass media in their coverage of environmental issues and problems which have been X-rayed in this study. Hardly any mass media house that knows its job will refuse an advertisement or an advertorial on the environment. Chevron Plc seems to be doing this now quite successfully, nationally and internationally i.e. using advertisements and advertorials to communicate to its publics what it has been doing and its policies in the area of environmental protection and sustainable development. So, we should not depend on journalistic outputs like news and editorials initiated by them alone, to communicate effectively on our companies' contributions, policies and activities on the environment.

This study has shown these outputs to be grossly inadequate, quantitatively and qualitatively. It has also revealed that the journalists do not seem to be properly or professionally motivated, equipped and prepared to give effective coverage to our organizations' roles and contributions in the area of environmental protection and sustainable development. The study went further to show that these journalists' performance seems to be the same, whether they work in government-owned or privately-owned media organizations.

The public relations managers however, need in their own interest, to get their organizations, when they are capable, to help organize seminars on environmental public relations and effective coverage of the environment for the journalists. This will help to further equip and sensitize them to do better than they have done in the area of covering environmental issues and problems.

Issues and Crisis Management

It has been shown by research that poorly managed issues often degenerate to crisis (Nwosu, 1996). Environmental issues are among the most delicate issues for many organizations in Nigeria. Public relations managers need therefore to sharpen their environmental scanning capabilities to be efficient and effective issues managers. This study has shown that they do not have to depend largely on media reports in their environmental scanning and environmental audits. They need to go far beyond the mass media to be successfull in this area.

Public Relations Research

Closely related to the above is the practical implication of this study for more and improved public relations research among Nigerian public relations managers. Many public and community relations managers in Nigeria still shy away from quantitative research. This study has shown that quantitative research is not only necessary in the management of companies' communications and activities on the environment, but also very necessary in all other areas of modern public relations practice. It is an example on its own of how one type of public relations research (media audit) can be designed and executed.

Government Relations Management

One implication of this study for the public relations manager is that it should serve as constant reminder to them that governments in Nigeria and everywhere (Federal, State and Local Governments) are always interested and have been always involved in environmental protection and sustainable development. Positively, this realization should help public relations managers to always remember that these governments will welcome joint venture or co-operative projects between them and their organizations or companies. They should therefore work out plans of action on the environment that will help to improve their relations with government and government agencies. The same can be said about our companies and NGO relations, as well as community relations. Any community will welcome such co-operative ventures on the environment and sustainable development, any day any time, to our companies' image management advantage.

On the negative side, public relations managers should deduce from this study, revelation of high government interest in the environment and sustainable development, that any neglect or poor handling of environmental issues and problems is likely to attract negative sanctions and perceptions of their companies by these governments. The same can also be said of NGOs, communities and their environmental activist organizations, when our companies perform badly in environmental issues and problems communications and management.

Conclusion

The road to R10 or achieving the objectives of the UN Agenda 21 on the environment may be far. But the present reality is that we and our organizations must start working harder than we have done before, to contribute continually towards the achievement of these objectives. We must give more respect to the environment and contribute towards the maintenance of the ecosystem. We must continue to contribute towards ensuring biodiversity, carry out periodic environmental audits and environmental impact assessments for our organizations, ensure regular environmental sanitation, contribute to poverty reduction, reduce environmental pollution, help to balance the equation between environment and development (Sustainable development) and do a lot more to be seen as good and responsible corporate citizens today and in the next millennium. This is the only way we can avoid crisis, conflicts and other problems, which will destroy us and our organizations, if we continue to abuse and destroy the environment.

This study has shown us that one of our allies in this onerous task of environmental protection and sustainable development, the mass media, cannot be depended entirely upon. It has brought the fact home to us that as responsible and responsive public relations professionals, we must have to find ways of remaining abreast of contemporary environmental issues and problems in order to help our organizations to contribute positively to the environmental protection and sustainable development crusade, in our own interest as homo sapiens, managers and professionals. A word is enough for the wise, as the saying goes. But we have deliberately used many words here to drive the message home and to handle the many "arms" and "legs" of these problems involved in communicating and managing environmental issues and problems effectively in contemporary society and from the point of view of the public relations manager.

Chapter Twelve

Case Study on Environmental Incidents and Public Relations Strategy

Introduction and problem statement.

Environmental incidents have been a major problem for the oil and gas industry in Nigeria for a long time. The nature, magnitude and frequency of these incidents have made them to remain a serious thorn in the flesh for both the oil and gas industry and the Nigerian government. They range from oil spillage and gas flaring to community hostilities and even international image crisis for the country. For example, the Ogoni crisis of 1993 dealt a major blow to the oil companies operating around Ogoni and to Nigeria's image abroad, resulting in the country's suspension from the Commonwealth (Nwosu, 200).

The above scenario make it imperative that managers of the Nigerian oil and gas industry must work closely with the government and the producing communities to find lasting solutions to the problems caused by these environmental incidents. And there is no doubt that Public Affairs Managers working in these oil and gas companies have the key to the building of and sustenance of mutual understanding among the parties involved through effective two-way communication and other public relations activities. These managers, in order to achieve the above stated objective, must endeavour to keep constantly abreast of

1. The key issues involved in the occurrence and management of these environmental incidents.

2. Know the best practices in the management of these environmental incidents and

3. Be able to recommend appropriate responses and actions by their companies to these incidents.

Towards the realization of the above three objectives, the following case study has been developed from a study that focused on the most critical of the environmental community hostilities and presented here as a basis for discussion and practice exercises. The research from which the practical case study in this chapter was developed for this book, was carried out by one of Professor Ikechukwu E. Nwosu's postgraduate students, H.G.O. Igben, under his (Nwosu's) close supervision and corrections. The very rigorous research lasted for more than one year and used a multi-methodological approach that is summarized at the end of this case study (Nwosu, 2000).

The Case Study

Community hostilities in the oil industry have resulted to huge financial loss to oil companies and government. The negative effect on an economy that is oil dependent for 90% foreign exchange earning must be very devastating. For example, in 1993 it was discovered that the Nation registered a loss of over N2.732 billion within nine months arising from disruptions of oil operations in Ogoni land, Rivers State; within the same year, cost of fixing up damages arising from 158 reported incidents of communities disturbances accounted for N300 million of the various oil companies funds (Igben, 1998).

At the root of the above drama is the dissenting cry of utter neglect by the oil producing communities which constantly alleged that their natural endowment which should have been extended to the development of their communities are being used lavishly for the development of communities other than theirs at their very expense. This trend as identified in this study, suggests a threat to the possible existence of good neighbourliness between the oil communities and oil producing companies. Politically, it poses a serious threat to the stability and unity of the country. And unless something urgent is done to address this situation, the Nations' economy through oil revenue will be much battered. It was discovered that the companies under study are substantially committed to the promotion of good neighbourliness. For instance, in 1992, the two companies under study spent over N250 million on community relations activities (Igben, 1998).

The Problems

Despite the above level of commitment, the oil communities under study do not accept this as adequate. This is evident from the escalating tempo of community hostilities incidents. While some host communities believe that amenities provided by the oil companies are sub-standard, others have a strong feeling that there is so much imbalance between what the oil companies realize from their land in relation to what they get in return (Igben, 1998).

The Key Findings

Government and Oil Industry Commitment

It was found that there is unrestrained commitment by the oil companies and government towards addressing identified negative feelings of the host communities. Revenue allocation, which was at 1.5% to the oil communities, was later reviewed upward to 3%. Another step in this direction was the establishment of the Oil Mineral Producing Areas Development Commission (OMPADEC) to accelerate the development of the oil producing communities through the judicious management of the 3%

allocation. Federal Environmental Protection Agency (FEPA) was established towards the protection of the environment from abuse by oil companies. The OMPADEC has been replaced by the NDDC.

Most recent on the list was the establishment of Niger Delta Environmental Protection Survey, which represents the effort of Shell Petroleum Development Company towards the ascertainment of the state of the ecosystem of the oil producing communities with view to mustering redressing measures (Igben, 1998).

Poor or Inadequate Needs Identification

Most community assistance projects are found to be out of touch with the reality of the specific needs of the oil producing communities. For instance, at Nembe, in Rivers State, the fishpond established for them seems to make no meaning to the people since they have the creeks to fish in Moffat (1995:18). Until the specific needs of the oil communities are identified, the level of goodwill and acceptance that the oil companies will enjoy in such communities will be either very low or non-existent (Igben, 1998).

Inadequate Management of Compensation Payment

Delay and non-payment of fair and adequate compensation for acquired or damaged third party property is fundamental to some community hostilities. Gatekeepers in the chain of interaction between the oil companies and oil communities are suspect in either failure to convince management towards prompt response or diversion of part of the funds to selfish ends.

Slow Response To Oil Spills

Also, delay in giving attention to oil spills has been responsible for some community hostilities. This study found that there were incidents of oil spills, which were caused by rusty pipeline in Nembe creeks, and Ahoada resulting in flame with the slightest contact with fires sparks causing loss of lives and properties. Green Peace International (Olukoya, 1995:15) believe that out of 100 countries where Shell operates 40% of its spills occur in Nigeria, which were attributed to out-dated materials.

Inadequate Communication

This study also discovered that lack of effective communication leading to misperception and suspicious at the grassroots is responsible for some community hostilities. The oil companies are responding fast with the adoption of people's parliament, open forum, needs assessment meeting and school talks, to mention but a few, towards bridging of the gap hitherto inherent in the relationship between the duo (Igben, 1998).

Inadequate Number of Community Liaison Officers

The host communities of 12 major ethnic groups were found to be suffering from the problem of inadequate coverage by community Liaison or Relations Officers. The community Liaison Officers made of about 28 are inadequate for the effective spread over 800 communities, the implication of which could mean lack of frequent understanding and acceptance (Igben, 1998).

Poor PR Professionalism

Another aspect is the caliber of manpower. The study revealed that many oil companies operating in Nigeria do not have professionals who are trained in the nitty-gritty or PR practice. Consequently, they find it difficult in monitoring and preventing trends that are provocative of crisis. For instance, PR professional would rather engage in meaningful peaceful dialogue, thoroughly exhaust this dimension, than, resorting to use of force at the first instance. The genesis of Ogoni crisis is a direct reflection of this conclusion (Igben, 1998).

Reactive Vs Proactive PR Practices

A good percentage of the Community Relations activities were reactive rather than proactive. The statistics showed that the oil companies most often embark on community assistance projects after they have had their activities disrupted (Igben, 1998).

Inadequate Training and Retraining

A larage percentage of the Community Relations Officers (CRO) do not have the privilege of going for regular sensitizing training programme. This of course, makes them to rely on residual knowledge. That is, those of social science extraction. But for those of non-allied PR discipline, they rely on their experiences which is often fraught with numerous mistakes and extra cost (Igben, 1998).

Host Communities Perception of Affluence as a Problem

The prevalent poor economic situation in the country juxtaposed against the background of affluent oil companies' staff is a causative factor to most of the community hostilities. Host communities see themselves as a people in extreme want in the midst of plenty (Igben, 1998).

Brief Note on the Methodology of this Case Study

The research for this case study used both primary and secondary data or information sources. It studied a sample of respondents taken from 9 oil companies and 15 communities in oil producing areas, which were used as the population for the study. A total of 138 respondents and 45 managers from the oil companies (including selected public affairs managers) were studied. The sample survey method of research was used with two structured questionnaires as test instruments, one for the selected management staff of the oil companies and one for selected member from the oil producing communities selected. Five different hypotheses were formulated and tested in the study (Igben, 1998).

The identities of the respondents, the oil companies and the communities, have been left out of this case study as part of the ethical requirements of case study preparation or writing (Nwosu, 1996).

Guiding Questions for the Case Study

What are the key issues involved in the occurrence and management of environmental incidents in the Nigerian oil and gas industries that can be deduced from this case study? List and discuss them.

1. What other key issues on environmental incidents can you think about (outside this case study) List and discuss them.

2. What PR best practices can you identify or deduce from this case study on the management of issues and crises in Nigerian organizations. List and discuss.

3. What other PR best practices can you recommend? List and discuss them.

4. Discuss the appropriateness or inappropriateness of the responses and actions of the oil and gas companies in Nigeria to environmental incidents in the oil producing communities.

5. Recommend appropriate responses and actions by the oil and gas companies to Environmental incidents in the oil producing communities.

6. React to all of the key findings reported in the case study above (i.e. 1-9).

7. List and discuss the implications of the findings reported in this study for improved public relations crisis and issues, management and practices in customs and excise.

8. What lessons should the customs and Excise learn from this case study on issues and crisis management? How can the PR Department and entire management ensure the application of these lessons wherever necessary? (Nwosu, 2000).

APPENDIX I

EXCERPT FROM NIGERIA'S NEW NATIONAL POLICY ON THE ENVIRONMENT, 1999

Introduction

Nigeria is committed to a national policy that will ensure sustainable development based on proper management of the environment. This demands positive and realistic planning that balances human needs against the carrying capacity of the environment. This requires that a number of complementary policies, strategies and management approaches are put in place which should ensure, among others, that:

Environmental concerns are integrated into major economic decision-making process;

Environmental remediation costs are built into major development projects;

Economic instruments are employed in the management of natural resources;

Environmentally friendly technologies are applied;

Environmental Impact Assessment is mandatory before any major development project is embarked upon;

Environmental monitoring and auditing of existing major development projects are routinely carried out;

This policy, in order to succeed, must be built on the following sustainable development principles;

The precautionary Principle which holds that where there are threats of serious or irreversible damage, the lack of full scientific knowledge shall not be used as a reason for postponing cost-effective means to prevent environmental degradation, Pollution Prevention Pays Principles (3P+) which encourages industry to invest positively to prevent pollution;

The Pollution Pays Principle (PPP) which suggests that the polluter should bear the cost of preventing and controlling;

The User Pays Principle (UPP), in which the cost of a resource to a user must include all the environmental costs associated with its extraction, transformation and use (including the costs of alternative or future uses forgone);

The Principle of Inter-generational equity which requires that the needs of the present generation are met without compromising the ability of future generations to meet their own needs; The Principle of Intra-generational Equity which requires that different groups of people within the country and within the present generation have the right to benefit equally from the exploitation of resources and that they have an equal right to a clean and healthy environment.

The Principle of Participation which requires that decisions should as much as possible, be made by communities affected or on their behalf by the authorities closest to them.

This new policy thrust is based on fundamental re-thinking and a clearer appreciation of the interdependent linkages among development processes, environmental factors as well as human and natural resources. Since development remains a national priority, it is recognized that the actions designed to increase the productivity of the society and meet the essential needs of the populace must be reconciled with environmental issues that had hitherto been neglected or not given sufficient attention.

In enunciating a national policy on the environment, cognizance must be taken of the various institutional settings and professional groupings, as well as the complex historical, social, cultural and legal considerations, which have been and continue to be involved in the identification and implementation of measures designed to solve national environmental problems. The provisions of the Policy have thus been informed by recent national policy initiatives in Science and Technology, Agriculture, Health, Industry, Oil and Gas, Population, Culture, etc as well as the major international efforts in the field of environment. The Policy aims to provide a national, practicable, coherent and comprehensive approach to the pursuit of economic and social development in a way that minimizes contradictions and duplications, while enhancing inter-and intra-sectoral co-operation and effectiveness at all levels.

Since the health and welfare of all Nigerians depend on making the transition to sustainable development as rapid as possible, this national Policy on the Environment provides the concepts and strategies which will lead to the procedures and other concrete actions required for launching Nigeria into an era of social justice, self-reliance and sustainable development as we enter the 21st Century.

Policy Goal

The goal of the National Policy on the Environment is to achieve sustainable development in Nigeria, and in particular to:

a. Secure a quality of environment adequate for good health and well-being;

b. Conserve and use the environment and natural resources for the benefit of present and future generations;

c. Restore, maintain and enhance the ecosystems and ecological processes essential for the functioning of the biosphere to preserve biological diversity and the principle of optimum sustainable yield in the use of living natural resources and ecosystems;

d. Raise public awareness and promote understanding of the essential linkages between the environment, resources and participation in environmental improvement efforts; and

e. Co-operate in good faith with other countries, international organization(s) and agencies to achieve optimal use of trans-boundary natural resources and effective prevention or abatement of trans-boundary environmental degradation.

Conceptual Framework

The National Policy on the Environment is basically a programme of actions rooted in a conceptual frame within which the linkages between environmental problems on the one hand and their causes, effects and solutions on the other hand can be discerned. This is achieved in the policy document through five major policy initiatives, viz:

a. Preventive activities directed at the social, economic and political origins of the environmental problems;

b. Abatement, remedial and restorative activities directed at the specific problems identified, and in particular. Problems arising from industrial production process; Problems caused by rapid population growth and the abundant excessive pressure of the population on the land and other resources; and problems due to rapid growth of urban centres;

c. Design and application of broad strategies for sustainable environmental protection and management at systematic or sub-systematic levels;

    d. Enactment of necessary legal instruments designed to strengthen the activities and strategies recommended by the POLICY;

    e. Establishment/emplacement of management organs, institutions and structures designed to achieve the policy objectives.

Strategies For Implementation

    The implementation of the National Policy on the environment depends on specific actions directed towards all sectors of the economy and problem areas of the environment. Consequently, the approach to problem solving adopted in this Policy is predicted on an integrated, holistic and systemic view of environmental issues.

    The action envisaged will establish and/or strengthen legal, institutional, regulatory, research, monitoring, evaluation, public information and other relevant mechanisms for ensuring the attainment of the specific goals and targets of the policy.

    It is also expected that these strategies will lead to:

a. Improvement in the quality of life of the people;

b. The establishment of adequate environmental standards as well as the monitoring and evaluation of changes in the environment and the adoption of appropriate restorative measures;

c. The acquisition and publication of up-to-date environmental data and the dissemination of relevant environmental information;

d. Prior environmental assessment of proposed activities which may impact the environment or the use of a natural resource.

e.

    Source: FEPA (1999), Revised National Policy on the Environment, pp. 1-5 (of 45 pages) (Abuja, Nigeria).

APPENDIX II

Decree No. 58

(30th December, 1988) commencement

THE FEDERAL MILITARY GOVERNMENT

Hereby decrees as follows:

PART 1-ESTABLISHMENT, MEMBERSHIP

FUNCTIONS AND POWERS OF THE FEDERAL

ENVIRONMENTAL PROTECTION AGENCY

1.   There is hereby established a body to be known as
the Federal Environmental Protection Agency (hereinafter in   establishment of the
this Decree referred to as "the Agency") which under that   Federal Environmental
name shall be a common seal, and may sue and be sued in its   Protection Agency
corporate name.

2.   –(1) The Agency shall consist of the following   Membership of
     members, that is to say –   the Agency
a.   Chairman who shall be a person with wide knowledge in environmental matters to be
     appointed by the President, Commander-in-Chief of the Armed Forces.

b.   Four distinguished scientists to be appointed by the President, Commander-in-Chief
     of the Armed Forces and one representative each from the following Federal
     Ministries-
(i)   Health;
(ii)   Science and Technology;
(iii) Works and Housing;
(iv)   Agriculture, Water Resources and Rural Development; Industries;
(v)   Mines, Power and Steel;
(vi)   Employment, Labour and Productivity;

(vii) Petroleum Resources (Petroleum Resources Department);

(viii) Transport;

(ix) Aviation; and

(x) The Director of the Agency

(2) Subject to section 3 of this Decree, a member who is not a public officer-

a. shall hold office for a period of four years from the date of his appointment and shall be eligible for reappointment for one further term of four year only;

b. shall be paid such remuneration and allowances as the President, Commander-in-Chief of the Armed Forces may, from time to time, determine.

(3) The supplementary provisions contained in the Schedule to this Decree shall have effect with respect to the proceedings of the Agency and the other matters therein mentioned.

Removed from   3.   The officer of a member who is not a public officer shall become
Office of   vacant if he resigns his office by a letter addressed by him to
member of   continue in office as a member in which case the Minister, or if
the Agency   the Minister is satisfied that it is not in the interest of the Agency
                   for the person appointed to continue in office as a member in
                   which case the Minster shall, with the approval of the President,
                   Commander-in-Chief of the Armed Forces, notify the Member
                   in writing to that effect.

Functions of   4.   The Agency shall, subject to this Decree, have responsibility for
Agency .   protection and development of the environment in for the general
                    and environmental technology, including initiation of policy in
                    relation to environmental research and technology; and without
                    prejudice to the generality of the foregoing, it shall be the duty
                    of the Agency to -

(a) advise the Federal Military Government on national environmental policies and priorities and on scientific and technological activities affecting the environment.

(b) prepare periodic master plans for the development of environmental science and technology and advise the Federal Military Government on the financial requirements for the implementation of such plans.

(c)   promote co-operation in environmental science and technology with similar bodies in other countries and with international bodies connected with the protection of the environment.

(d)   co-operate with Federal and State Ministries, Local Government Councils, statutory bodies and research agencies on matters and facilities relating to enviornemntal protection; and

(e)   to carry out such other activities as are necessary or expedient for the full discharge of the functions of the Agency under this Decree.

4. In carrying out the functions prescribed on section 4 of this Decree Powers of the and Powers in other provisions of this Decree, it shall be lawful for the

Agency to give

Agency to-grants, etc

(a)   make grants to suitable authorities and bodies with similar functions for demonstration and for such other purposes as may be determined appropriate to further the purposes and provisions of this Decrees;

(b)   collect and make available, through publications and other appropriate means and in co-operation with public or private organizations, basic scientific data and other information pertaining to pollution and environmental protection matters;

(c)   enter into contracts with public or private organizations and individuals for the purpose of executing and fulfilling its functions and responsibilities pursuant to this Decree;

(d)   establish, encourage and promote training programmes for its staff and other appropriate individuals from public or private organizations;

(e)   enter into agreement with public or private organizations and individuals to develop, utilize, co-ordinate and share environmental monitoring programmes, research effects, basic data on chemical, physical and biological effects of various activities on the environment and other environmentally related activities as appropriate;

(f) establish advisory bodies composed of administrative, technical or other experts in such environmental areas as the Agency may consider useful and appropriate to assist it in carrying out the purposes and provision of this Decree.

(g) establish such environmental criteria, guidelines, specifications or standards for the protection of the nation's air and inter-state waters as may be necessary to protect the health and welfare of the population from environmental degradation;

(h) establish such procedure for industrial or agricultural activities in order to minimize damage to the environment from such activities;

(i) maintain a programme of technical assistance to bodies (public or private) concerning implementation of environmental criteria, guidelines, regulations and standards and monitoring enforcement of the regulations and standards thereof; and

(j) develop and promote such processes, methods, devices and materials as may be useful or incidental in carrying out the purposes and provisions of this decree.

Power of the 6. Minister to give agency directions	Subject to the Decree, the Minister may give to the Agency directions of a general nature or relating generally to particular Matters, but not to any particular individual or case with regard to the performances by the Agency of its functions under this Decree and it shall be the duty of the Agency to company with the directions.
Director and other staff of the Agency	7. – (1) There shall be appointed by the President, Commander-in-Chief of the Armed Forces a Director of the Agency
	(2) The Director who shall be the chief executive of the Agency shall hold office on such terms and conditions as may be specified in his letter of appointment and, subject thereto, on such other terms and conditions as may be determined by the Agency in respect of the other employees thereof.

(3) The Agency may appoint such other persons to be employees of the Agency as it may deem fit.

(4) The remuneration and tenure of office of employees (other than the Director) shall be determined by the Agency with the approval of the Minister.

8. – (1) It is hereby declared that service in the Agency shall be public service for the purposes of the Pensions Act 1979 and accordingly officers and other staff of the Agency shall in respect of their services be entitled to such pensions. Gratuities and other retirement benefits as are prescribed hereunder.

(2) For the purposes of the application of the provisions of the Pensions Act 1979, any powers exercisable thereunder by a Minister or other authority of the Government of the Federation (not being the Power to make regulations under section 23 thereof) are hereby vested in and shall be exercisable by the Agency and not by any other person or authority.

9. The Director shall, subject to the policies laid down by the Agency, develop programmes to carry out the purposes and provisions of this Decree and, without prejudice to the generality of the foregoing, shall, in particular and in consultation with appropriate Agencies-

Power of
Director

(a) establish programmes for the prevention, reduction and elimination of pollution of the nation's air, land and inter-State waters, as well as national programmes for restoration and enhancement of the nation's environment;

(b) encourage and promote the co-ordination of environmentally, related activities at all levels;

(c) utilize and promote the expansion of research, experiments, surveys and studies by public or private agencies, institutions and organizations concerning causes, effects, extent, prevention, reduction and elimination of pollution and such other matters related to environmental protection as the Agency may, from time to time, determine necessary and useful; and

(d) conduct public investigations on pollution.

Power to
accept gift

10. – (1) Subject to subsection (2) of this section, the Agency may accept gift of land, money, books or other property upon such terms and conditions, if any, as may be specified by the person making the gift.

(2)   The Agency shall not accept any gift if the conditions attached thereto by the persons making the gift are inconsistent with the functions of the Agency under this Decree.

Residence,   Offices
and premises

11. – (1) For the purposes of providing residential accommodation for its staff, offices and premises as may be considered necessary for the performance of its functions under this Decree, the Agency may –

purchase or take on lease any interest in land; and

build, furnish, equip and maintain residential quarters, offices and premises.

1978 No. 6

(2)   Subject to the Land Use Act 1978, the Agency may, with the approval of the Minister, sell or lease any residential quarters, land, offices or premises held by it and no longer required for the performance of its functions.

Financial provision

12. – (1) The Agency shall establish and maintain a fund from which shall be Defrayed all expenses incurred by the Agency.

(2)   The Agency shall cause to be kept proper accounts established under subsection.

(1)   of this section -

such sums as may, from time to time, be granted to the Agency by the Federal Military Government;

all moneys raised for the purposes of the Agency by way of gifts, grants in aid, testamentary dispositions and sales of publications;

all subscriptions, fees and charges for services rendered by the Agency and all other sums that may accrue to the Agency from any source.

13. – (1) The Agency may, with the consent of the Minister or in minister or in accordance with the general authority given by the Federal Military Government, borrow by way of loan or overdraft from any source any moneys required by the Agency to meet its obligations and its functions under this Decree, so however that no such consent or authority shall be required where the sum or the aggregate of the sums involved at any time does not exceed such amount as it for the time being specified in relation to the Agency by the Federal Military Government

borrowing power

(2)   The Agency may, subject to the provisions of this Decree and the conditions of any trust created in respect of any property, invest all or any of its funds with the like consent or general authority.

(3)   The Agency may invest any of its surplus funds in such securities as may be permitted by law.

14. – (1) The Agency shall cause to be prepared not later than 6 months before the end of each year an estimate of the expenditure and income of the Agency during next succeeding financial year and when prepared they shall be submitted to the Minister-

Annual than estimate accounts

(2)   The Agency shall cause to be kept proper accounts and proper records in relation thereto and when certified by the Agency such accounts shall be audited as provided in subsection (3) of this section.

(3)   The accounts of the Agency shall be audited as soon as may be practicable After the end of each year by auditors appointed from the list and in accordance with the guidelines supplied by the Auditor-General of the Federation.

PART II – (1) NATIONAL ENVIRONMENTAL STANDARDS
Water quality

Federal quality standards

15. – (1) The Agency shall make recommendations to the Ministers for the quality purpose of establishing water for the inter-state waters of standards Nigeria to protect the public health of welfare and enhance the quality of water to serve the purpose of this Decree.

(2)   In establishing such standards, the Agency shall take into consideration the use and value for public water supplies, propagation of fish and wildlife, recreational purposes, agricultural, industrial and other legitimate uses.

(3)   The Agency shall establish different water quality standards for different uses.

Effluence Limitation

16. – (1) The Agency shall, as soon as possible after the commencement of this Decree, establish effluent limitations for new point sources which shall require application of the best control technology currently available and implementation of the best management practices.

(2)   The Agency shall, as soon as possible after the commencement of this Decree, establish effluent limitations for existing point sources which shall require the application of the best management practices under circumstances as determined by the agency, and shall include schedules of compliance for installation and operation of the best practicable control technology as determined by the Agency.

Air quality, etc. 17. – (1) The Agency shall establish more criteria, guidelines, specifications and standards to protect and enhance the quality of Nigeria's air resources so as to promote the public health or welfare and the normal development and productive capacity of the nation's human, animal or plant life, and include in particular –

(a)   minimum essential air quality standards for human, animal or plant health;

(b)   the control of concentration of substances in the air, which separately or in combination are likely to result in damage or deterioration of property, or of human, animal or plant health;

(c)   the most appropriate means to prevent and combat various forms of atmospheric pollution;

(d)   controls for atmospheric pollution originating from energy sources, including that produced by aircraft and other self-propelled vehicles and in factories and power generating stations;

(e)   standards applicable to emission from any new mobile source which in the Agency's judgment causes or contributes to air pollution which may reasonably be anticipated to endanger public means to reduce emission to permissible levels;

(f)   the Agency may establish monitoring stations of networks to locate sources of atmospheric pollution and determine their actual or potential danger.

18. – (1) The Agency shall undertake to study data and recognize developments in international force and other countries regarding the cumulative effect of all substances, practices, processes and activities which may affect the stratosphere, especially ozone in the stratosphere. The Agency may make recommendations and programmes for the control of any substance, practice, process or activity which may reasonably be anticipated to affect the stratosphere, especially ozone in the stratosphere, when such effect may reasonably be anticipated to endanger public health or welfare.
Ozone protection

(3)   For purpose of this section, "stratosphere" means that part of the atmosphere above the troposphere.

Noise control. 19. – (1) The Agency shall, as soon as practicable after the commencement of this Decree, in consultation with appropriate authorities –

(a) identify major noise sources noise criteria and noise control technology; and

(b) establish such noise abatement programmes and noise emission standards as it may determine necessary to preserve and maintain public health or welfare.

(2) Any noise criteria identified under this section shall reflect the scientific knowledge most useful in indicating the kind and extent of all identifiable effects on the public health or welfare, which may b e expected from differing qualities and quantities of noise.

(4) The Agency shall make recommendations to control noise originating from industrial, commercial, domestic, sports, recreational, transportation of other similar activities.

Hazardous substances, etc

Discharge of hazardous substance and related offences

20. – (1) The discharge in such harmful qualities of any hazardous substances into the air or upon the land and the waters of Nigeria or substances and the joining shorelines is prohibited, except where such discharge related is permitted or authorized under any law in force in Nigeria offences.

(2) Any person who violate the provisions of subsection (1) of this section commits and offence and shall on conviction be liable to a fine not exceeding N100,000.00 or to imprisonment for a term not exceeding 10 years or to both such fine and imprisonment.

(3) Where an offence under subsection (1) of this section is committed by a body corporate it shall on conviction be liable to a fine not exceeding N500,000.00 and an additional fine of N1,000.00 for every day the offence subsists.

(4) Where any offence under this Decree has been committed by a body corporate, the body corporate and every person who at the time the offence was committed was in charge of, or was responsible to the body corporate for the conduct of the business of the

body corporate shall be deemed to the guilty of such offence and shall be liable to be proceeded against and punished accordingly.

(5)   The Agency shall, as soon as possible after the commencement of this Decree, determine for the purposes of this section what substances are hazardous substances and such hazardous substances the discharge of which shall be harmful under the circumstances to public health or welfare and, for this purpose, the Agency shall take into account such special circumstances including locations, quantity and climatic conditions relating to discharges as it may determine appropriate.

(6)   Notwithstanding the provisions of this section or of any other sections of this Decree, the provisions of the Harmful Waste (Special Criminal Provisions, Etc) Decree 1998 shall apply in respect of any hazardous substance constituting harmful waste as defined in section 15 thereof.

21. – (1) Except where an owner or operator can prove that a discharge was Spiller's caused solely by a natural disaster or an act of war or by sabotage, such owner liability, or operator of any vessel or onshore or offshore facility from which the hazardous substance is discharged is in violation of section 20 of this Decree, shall in addition to the penalty specified in that section be liable for –

(a)   the cost of removal thereof, including any costs which may be incurred by any Government body or agency in the restoration or replacement of natural resources damaged or destroyed as a result of the discharge; and

(b)   costs of third parties in the form of reparation, restoration, restitution or compensation as may be determined by the Agency from time to time.

(2)   The owner or operator of a vessel or onshore or offshore facility from which there is a discharge in violation of section 20 of this Decree shall, to the fullest extent possible, act to mitigate the damage by –

(a)   giving immediate notice of the discharge to the Agency and any other relevant agencies.

(b)   beginning immediate clean-up operations following the best available clean-up practice and removal methods as may be prescribed by regulations made under section 22 of this Decree; and

(c)   promptly complying with such other directions as the Agency may from time to time, prescribed.

Removal 22. methods, etc	The Minister for purposes of this part of this Decree may, by regulations, prescribe any specific removal methods, national contingency plans, financial responsibility levels for owners or operators of vessels, or onshore or offshore facilities, notice and reporting requirements, penalties, and compensation as he may determine necessary to minimize pollution by any hazardous substance
Co-operation with the Ministry of Petroleum Resources	23. The Agency Shall Co-Operate With The Ministry Of Petroleum Resources. (Petroleum Resources Department) for the removal of oil of related pollutants discharged into the Nigerian environment and play such supportive role as the Ministry of Petroleum Resources (Petroleum Resources Development) may from time to time request from the Agency)

PART III –   ESTABLISHMENT OF STATE AND LOCAL GOVERNMENT
ENVIRONMENTAL PROTECTION BODIES

Establishment of State and Local government bodies	24. – the Minister shall, as soon as possible after the commencement of this Decree, encourage States and Local Government Councils to set up their own local Environmental Protection Bodies for the purposes of maintaining good environmental quality in the areas of related pollutants under their control subject to the provisions of this Decree.

PART IV – SUPPLEMENTARY AND MISCELLANEOUS

Enforcement Powers

25.   For the purposes of enforcing this Decree, any authorized officer may, without a warrant: -

Powers to inspect etc.

(a) require to be produced, examine and take copies of, any license, permit, certification or other document required under this Decree or any regulations made there under;

(b) require to be produced and examine any appliance, device or other item used in relation to environmental protection

26. – (1) Any authorized officer, where he has powers to reasonable grounds for believing that an offence has been committed against this Decree or any regulations made there-under, may without a warrant -

Powers to search, seize and arrest

(a)   enter and search any land, building, vehicle, tent, vessel, floating craft or any inland water or other structure whatsoever, in which he has reason to believe that an offence against this Decree or any regulations made thereunder has been committed;

(b)   perform tests and take sample of any substances relating to the offence which are found on the land, building, vehicle, tent, vessel, floating craft or any inland water or other structure whatsoever, searched pursuant to paragraph (a) of this subsection;

(c)   cause to be arrested any person who he has reason to believe had committed such offence; and

(a) Seizes any item or substance which he has reason to believe has been used in the commission of such offence or in respect of which the offence has been committed.

(2)   A written receipt shall be given for any article or thing seized under subsection (1) of this section and the grounds for such seizure shall be stated on such receipt.

Obstruction of Authorized

27. Any Person Who –

(a) Willfully Obstructs Any Authorized Officer In The Exercise Of Any Of The Officers Powers Conferred On Him By This Decree; Or

(b) Fails to comply with any lawful enquiry or requirements made by any authorized officer in accordance with the provisions of section 25 of this Decree. Commits an offence and shall on conviction be liable to a fine not exceeding N500,000.00 or to imprisonment for a term not exceeding 10 years or to both such time and imprisonment

Authorized Officers to disclose identify

28. – (1) Any authorize officer, not in uniform when acting under the provision of this Decree, shall, on demand, declare his office and produce to any person against whom he is taking action such identification or written authority as may reasonably be sufficient to show that he is an authorized officer for the purposes of this Decree.

(2) It shall not be an offence for any person to refuse to comply with any request, demand or order made by any authorized officer not in uniform, if such person fails to declare his office or produce such identification or written authority.

Procedures in of suits

29. – (1) No suit against the Agency, a member of the Agency or any respect employee of the Agency for any act done in pursuance or execution of any law, or of any public duties, or in respect of any alleged neglect or default in the execution of such law, duties or authority, shall lie or be instituted in any court unless it is commenced within twelve months next after the act, neglect or default complained of or, in the case of a continuance of damage or injury within twelve months next after the ceasing thereof.

(2) No suit shall be commenced against the Agency before the expiration of a period of one month after written notice of intention to commence the suit shall have been served upon the Agency by the intending plaintiff or his agent; and the notice shall clearly explicitly state – the cause of action;

(a) the particulars of the claim;
(b) the name and place of abode of the intending plaintiff; and

(c) the relief which he claims.

30. The notice referred to in section 29 (2) of this Decree and any summons, notice or other required or authorized to be served upon the Agency under the provisions of this Decree or any other law may be served by delivering the same to the Chairman or the Director, or by sending it by registered post addressed to the Director at the Secretariat of the Agency

service of documents

31. In any action or suit against the Agency no execution or attachment or process in the nature thereof shall be issued against the Agency but any sums of money which by judgment of the court is awarded against the Agency shall, subject to any directions given by the Agency, be paid from the general preserve fund of the Agency

Restriction on execution against the property of Agency.

32. Every member of the Agency, agent, auditor or employee for the time being of the Agency shall be indemnified out of the assets of the Agency against any liability incurred by him in defending any proceeding whether civil or criminal, if any such proceeding is brought against him in capacity as such member, agent, auditor or employee as aforesaid.

Indemnity of members of the Agency

33. The Agency shall, not later than 30th September in each year submit to the Minister a report on the activities of the Agency and its administration during the immediately preceding year and shall include in such report the audited accounts of the Agency.

Annual report

34. – (1) If a person knowingly or recklessly makes any statement in purported compliance with a requirement to furnish information which is false in a material particular, he commits an offence and shall on conviction and line not exceeding N200 or imprisonment for a term not exceeding 1 year or to both such fine and imprisonment

Material misrepresentation and impersonation

(2) Any person who falsely represents himself to be an authorized officer of the Agency and assumes to do any act or to attend in any place for the purpose of doing any act on behalf of the Agency shall be guilty of an offence under this Decree and on conviction shall be liable to imprisonment for a term not exceeding 2 years.

General penalties

35. Any person who contravenes any provisions of this Decree of any regulation made thereunder commits an offence and shall on conviction, where no specific penalty is prescribed therefore, be liable to a fine not exceeding N20,000 or to imprisonment for a term not exceeding 2 years or to both such fine and imprisonment.

Companies and Firms liable

36. Where any offence against this Decree or any regulations made thereunder has been committed by a body corporate or by a member of a partnership or other firm or business, every director or officer of that body corporate or any member of the partnership or other person concerned with the management of such firm or business shall, on conviction, be liable to a fine of not exceeding N500,000 for such offence and in addition shall be direct to pay compensation for any damage resulting from such breach thereof or to repair and restore the polluted environmental area to an acceptable level as approved by the Agency unless he proves to the satisfaction of the court that –

(a) he used due diligence to secure compliance with this Decree; and

(b) such offence was committed without his knowledge, consent or connivance

Miscellaneous

Power to make regulations

37. The Minister, on the advice of the Agency, may make generally for the purposes of this Decree but, without prejudice to the generality of the forgoing, the Minister may, in particular prescribe standards for –

(a) water quality;

(b) effluent limitation;

(c) air quality;

(d) atmospheric protection;

(e) ozone protection;

(f) noise control; and

(g) control of hazardous substances and removal methods.

38. In this Decree, unless the context otherwise Requires –   Interpretation
"appropriate agencies" means any government agencies which have
jurisdiction over the land or water affected by the pollution or any
government agencies which ordinarily have jurisdiction over the
operation which led to the pollution;

"authorized officer" means any employee of the Agency, any police officer not below the
rank of the Inspector of Police, or any customs officers;

"court" means the Federal High Court;

"Director" means the Director of the Federal Environmental Protection Agency;

"Disposal" includes both land-based disposal and dumping in waters and airspace of
Nigeria;

"Effluent limitation" means any restriction established by the Agency on quantities, rates
and concentration of chemical, physical, biological or other constituents, which are
discharge from point sources into the waters of Nigeria;

"Environment" includes water, air, land and all plants and human beings or animals living
therein and the inter-relationships, which exist among these or any of them;

"Hazardous substance" includes any substance designated as such by the Minister by order
published in the Gazetter;

"Minister" means the Minister charged with responsibility for the Environment

"New source" means any source, the construction of which commenced after the
publication of any regulations prescribing a standard of performance under this Decree,
which is applicable to such source;

"Offshore facility" means any facility, (including but not limited to motor vehicles and rolling stock) of any kind located over, in; on or under any land within Nigeria other than submerged land;

"Owner" or "operator" means –

(a) in the case of a vessel, any person owing, operating or chartering by demise such vessel;

(b) in the case of an onshore facility, the person owned or operated such facility immediately prior to such abandonment;

"Point source" means any discernible, confined and discrete conveyance, including but not limited to any pipe, ditch, channel, tunnel, conduit, well, discrete fissure, container, rolling stock, concentrated animal feeding operation or vessel or other floating craft form which pollutants are or may be discharged;

"Pollution" means man-made or man-aided alternation of chemical, physical or biological quality of the environment to the extent that is detrimental to that environment or beyond acceptable limits and "pollutant" shall be construed accordingly;

"Removal" means removal of hazardous substances from waters of Nigeria, including shorelines or the taking such other action as may be necessary to minimize or mitigate damage to the public health or welfare, ecology and natural resources of Nigeria;

"Waters of Nigeria" means all water resources in any form, including atmospheric, surface and sub-surface, and underground water resources where the water resources are inter-State, or in the Federal Capital Territory, territorial waters, Exclusive Economic Zone or in any other area under the jurisdiction of Federal Government.

39. This Decree may be cited as the Federal Environmental Protection Agency Citation Decree 1988.

ENVIRONMENTAL PUBLIC RELATIONS MANAGEMENT

SUPPLEMENTARY SCHEDULE SECTION 1(3)
SUPPLEMENTARY PROVISIONS RELATION TO THE AGENCY

Proceedings

1.   subject to this Decree and section 36 of the Interpretation Act 1964 (which provides for decision of a statutory body to be taken by a majority of its members and for the person presiding to have a second or casting vote), the Agency may make standing orders regulating the proceedings of the Agency or any committee thereof.

2.   Every meeting of the Agency shall be presided over by the Chairman or in his absence the members present at the meeting shall elect one of their members to preside at the meeting.

3.   The quorum at a meeting of the Agency shall consist of the Chairman (or in an appropriate case the person presiding at the meeting pursuant to paragraph 2 of this Schedule) and six other members.

4.   Where upon any special occasion the Agency desires to obtain the advice of any person on any particular matter, the Agency may co-opt that person to be a member for as many meetings as may be necessary; and that person while so co-opted shall have all the rights and privileges of a member except that he shall not be entitled to vote.

Committee

5.   – (1) Subject to its standing orders, the Agency may appoint such number of standing and ad hoc committees as it thinks fit to consider and report on any matter with which the Agency is concerned.

2.Every committee appointed under the foregoing provisions of this paragraph shall be presided over by a member of the Agency and shall be made up of such number of other persons, not necessarily members of the Agency, as the Agency may determine in each case.

3.The quorum of any committee set up by the Agency shall be as may be determined by the Agency.

6.   Where standing orders made pursuant to paragraph 1 of this Schedule provide for a committee of the Agency to consist of co-opted persons who are not members of the Agency, the committee may advise the Agency on any matter referred to it by the Agency and the members thereof may attend any meeting of the Agency for that purpose.

Miscellaneous

7.   The fixing of the seal of the Agency shall be authenticated by the signature of the Chairman and of the Director of the Agency.

8.   Any contract or instrument, which, if made by a person not being a body corporate, would not be required to be under seal, may be made or executed on behalf of the agency by the Director or by any other person generally or specially authorized to act for that purpose by the Agency.

9.   Any document purporting to be a contract, instrument or other document duly signed or sealed on behalf of the Agency shall be received in evidence and shall, unless the contrary is proved, be presumed without further proof to have been so signed or sealed.

10.   The validity of any proceedings of the Agency or of a committee thereof shall not be adversely affected –
(a) by any vacancy in the membership of the Agency or any committee thereof; or
(b) by any defect in the appointment of a member of the Agency or any committee thereof.

11.   Any member of the Agency or a committee thereof who has a personal interest in any contract or arrangement entered into or proposed to be considered by the Agency or committee thereof shall forthwith disclose his interest to the Agency

or the committee and shall not vote on any question relating to the contract or arrangement.

12.   No member of the Agency shall be personally liable for any act done or omission made in good faith while engaged on the business of the Agency.

MADE at Lagos this 30th day of December 1988.

GENERAL I.B BABANGIDA
President, Commander-in-Chief of the Armed Forces,
Federal Republic of Nigeria

EXPLANATION

(This note does not form part of the above Decree but is intended to explain its purport)

The Decree establishes the Federal Environmental Protection Agency with the following functions –

(a)   responsibility for monitoring and helping to enforce environmental protection measures;

(b)   co-operation with Federal and State Ministries, Local Government Councils, statutory bodies, and research agencies on matters and facilities relating to environmental protection.

For the effective implementation of its functions, the Decree confers additional powers on the Agency including the power to establish standards in certain environmental areas. In order to make the work of the Agency effective, the Decree confers upon its employees certain specific powers, for instance, power to inspect, search and seized and the power to arrest offenders.

PUBLISHED BY AUTHORITY OF THE FEDERAL MILITARY GOVERNMENT OF NIGERIA AND PRINTED BY THE MINISTRY OF INFORMATION AND CULTURE PRINTING DIVISION, LAGOS

FEDERAL ENVIRONMENTAL PROTECTION AGENCY AC, CAP 131, LFN, 1999 (as amended) by DECREE NO. 59 OF 1992.

Decree No. 59

ARRANGEMENT OF SECTIONS

SECTION

PART I. – ESTABLISHMENT, GOVERNING COUNCIL, FUNCTIONS AND POWERS OF THE FEDERAL ENVIRONMENTAL PROTECTION AGENCY

PART II. NATIONAL ENVIRONMENTAL STANDARDS

Water quality

Air quality and atmospheric protection

17.   Air quality, etc

18.   Ozone protection

Noise

19.   Noise control

Hazardous substances

20.   Discharge of hazardous substances

21.   Spiller's liability

22.   Removal methods, etc

23.   Co-operation with Federal Ministry of Petroleum Resources Department

APPENDIX III

NATIONAL ENVIRONMENTAL PROTECTION MANAGEMENT OF SOLID AND HAZARDOUS WASTES REGULATIONS, 1991

National Environmental Protection (Management of Solid and Hazardous Wastes) Regulations 1991

Commencement: 30th December, 1991

In the exercise of the powers conferred upon me by section 37 of the Federal Environmental Protection Agency Act and all other powers enabling that behalf, I Major General Mamman Tsofo Kontagora (Rtd) on the advise of the Federal Environmental Protection Agency, hereby make the following regulations:

PART 1 – OBJECTIVES AND DESIGNATION OF DANGEROUS WASTE

1.   The objectives of solid and hazardous waste management shall be to –

      (a)   identity solid, toxic and extremely hazardous   "Objectives of
             wastes dangerous to public health and   management of
             environment;   solid and
                                              hazardous waste

      (b)   provides for surveillance and monitoring of dangerous and extremely hazardous wastes and substances until they are detoxified and safely disposed of;

      (c)   provide guidelines necessary to establish a system of proper record keeping, sampling and labeling of dangerous and extremely hazardous waste;

      (d)   establish suitable and provide necessary requirements to facilitate the disposal of hazardous wastes;

(e)   research into possible reuse and recycling of hazardous wastes.

2.   – (1) All industries shall inform the Agency of all toxic   Functions of
           hazardous and radioactive substances which they keep   appropriate
           in their premises and/or which they discharge during   Governmental
           their production process   Agencies

(2)   The Agency shall maintain an up to date register of all industries which keep toxic, hazardous and eradicative substances or discharge toxic, hazardous and radioactive wastes.

(3)   The Agency shall prescribe to the relevant industries, factories or other institutions methods of controlling the generation of toxic, hazardous and radioactive wastes.

(4)   The Agency shall maintain a register of BANNED toxic, hazardous and radioactive substances and inform relevant industries, factories or institutions of the substances.

(5)   For the purpose of the disposal of toxic solid and hazardous wastes, the Agency shall –

           (a)   monitor and ensure that industries, factories or other institutions which discharge toxic, hazardous and radioactive waste as listed in column of Schedule 1 to these regulations, shall treat such hazardous wastes in the manner prescribed in Schedule 2 to these regulations;

           (b)   request from any industry on its register, information relating to the generation, handling, disposal of toxic hazardous and radioactive wastes; it shall be unlawful for any industry to withhold any such information;

           (c)   employ scientific and human resources to monitor and control all phases of life cycle of all substances likely to have an adverse effect on human health and environment;

(d)   determine and use the most advanced technology available for the disposal of toxic, hazardous and radioactive wastes;

(e)   set up regional bodies or committees to serve as "DUMP WATCH" for transboundary movement of toxic, hazardous and radioactive waste. The Agency shall prescribe necessary guidelines to the committees set up pursuant to paragraph (e) of these Regulations.

3.   A solid waste shall be determined to be a dangerous waste or extremely hazardous waste if it conforms with the provision of regulation 6 of these Regulations and it is out of waste specified as dangerous in the list of dangerous wastes specified in Schedule 6 to these Regulations.

PART II – DANGEROUS WASTE LIST

5.   A waste shall be designated as discharged chemical product, if it is handled in any of the manners described in Schedule 4 to these Regulations.

Dangerous Waste   6. – (1) waste shall be deemed to be dangerous waste if the waste
Source   appears in the list of dangerous waste contained in FAC – 000-000-9903 as listed in Schedule 12 to these Regulations.

(2)   A waste shall be regarded as dangerous if it is –

(a)   waste which appears in the dangerous waste source list as listed in Schedule 12 to these Regulations;

(b)   waste which is a residue from the management of a waste source list in Schedule 12 to these Regulations and identified as "D.W.", and

(c)   is described in the Footnotes of FACT – 000-000-9904 as an extremely hazardous waste in the hazardous and dangerous waste sources list in Schedule 13 to these Regulations.

(3)   Any waste appearing in the dangerous waste list shall be designated as exclusively hazardous waste (EHW).

Dangerous Waste   7. Infectious dangerous wastes shall include but not be limited to
Mixtures   infectious waste specified in Schedule 5 to these Regulations

8. – (1) Any waste whose constituents and concentration are known and which has not been designated as –

    (a)   a discarded chemical product;

    (b)   an infectious dangerous waste;

    (c)   a dangerous waste source, shall be deemed as a dangerous waste mixture which the provisions of these Regulations shall apply.

    (2)   A dangerous waste mixture shall also be determined as dangerous if –

        (a)   the category or degree of toxicity for each known constituent in the waste is known, or

        (b)   each known constituent in the waste is a halogenated hydrocarbon or a polycyclic aromatic hydrocarbon with greater than three rings and less than seven rings; or

        (c)   each known constituent of the waste is regarded by the International Agency for Research on Cancer (IARC) as human or animal positive or a suspected carcinogen.

    (3)   Any person who has dangerous waste materials shall use data available to him to determine the extent of toxicity, the person concerned shall apply to the Agency or its employees to determine whether the waste is contained in the Exclusive List of Registered Dangerous Substances in the Register with the Agency.

9. – (1) The toxic category for each toxic constituent in a waste shall   Determine toxicity
be determined by referring to the FEPA Register and by checking   in waste
this data against Schedule 6 to these Regulations.

    (2)   Where the toxic constituent classified under more than one of the four toxicity categories (Aquatic, oral, Inhalation or Dermal), the toxic constituent shall be assigned to the most acutely toxic category represented.

    (3)   The category of toxicity in a waste shall be determined in accordance prescribed in Schedule 6 to these Regulations.

10. – (1) The degree of toxicity shall be categorized according to the formular prescribed in Schedule 6 to these Regulations

categorization of toxic waste

(2)   If a person has established the toxicity of his waste by means of Bioassay test methods and has determined his waste toxicity's range, then he shall designate his waste according to the Toxic Dangerous Waste Designation in Schedule 7 to these Regulations.

11. – (1) Waste which contain halogenated hydrocarbon (HH) and/or polycyclic aromatic hydrocarbons with more than three rings and less than seven rings (PAH) shall be determined by the procedure specified in these Regulations

Persistent hazardous dangerous waste

(2)   A person shall – determine the concentration of (HH) and/or (PAH) in his waste by either testing his waste and specified in (a) of this sub-paragraph or by the calculation procedures described in (a) of this subparagraph, that is –

(a)   CONCENTRATION tests: A persons shall test his waste to determine its concentration level as stated in sampling and testing method below:

(b)   concentration calculation: If a person can demonstrate to FEPA beyond a reasonable doubt that any remaining persistent constituents for which he does not know the concentrations of would not contribute significantly to the total persistent concentration of his waste then he may calculate this concentration as follows:

(3)   A person whose waste contains one or more halogenated hydrocarbons for which the concentrations are known shall determine his total halogenated hydrocarbon concentration by summing the concentration percentages for all of his waste's significant halogenated hydrocarbons

Carcinogenic Dangerous wastes

12. – (1) A Substance Listed As An IARC (International Agency For Research On Cancer) Human or animal positive or suspected Carcinogenic and is an inorganic, respiratory, carcinogen shall be a carcinogenic substance provided it occurs in a triable format (that

is, if it is a waste which easily crumbles and forms dust which can be inhaled.

(2) Any person whose waste contains one or more IARC carcinogen(s) and it –

(a) the monthly or hatch waste quality exceeds 100kg or

(b) the concentration of any one positive (human or animal) carcinogen exceeds 1.0 per cent of the waste quantity

(i) shall designated such waste as EHW (and such designation shall be determined by (b) (ii), (iii);

(ii) the concentration of any one IARC positive (human or animal) carcinogen exceeds 0.01 per cent of the e waste quantity, such waste shall be designated DW; or

(iii) the total concentration summed for all IARC positive and suspected (human or animal) carcinogens exceeds (11.0 per cent of the waste quantity such waste shall be designated DW.

(c) a substance shall not be carcinogenic if it is rated as IARC human or animal positive or suspected carcinogen merely because of studies involving implantation of the substance into the animals as site cause for the IARC rating

Characteristic to determine solid waste as a dangerous waste

13. A solid waste shall be regarded as dangerous waste by any of the following parameters.

(a) ignitability;

(b) corrosivity;

(c) reactivity;

(d) extraction procedure toxicity (EPTOX);

(e) halogenated hydrocarbon concentration;

(f) polycyclic aromatic hydrocarbon concentration (PAH);

(g) static acute fish toxicity test;

(h) acute oral rate toxicity test;

(i) polychlorinated Dibenzo P-dioxins and dibenzofurans concentrations;

(j)   polychlorinate Biphenyls (PCB'S)

Methods and Test to determine representative samples

14. – (1) The Method Used For Obtaining Representative Samples Of A Waste Shall Vary With The Type And Form Of The Waste.

(2) The Agency shall consider such representative samples using any of the sampling method.

(3) If the waste samples have properties similar to the characteristic mentioned in regulation 13 of these Regulations and the materials indicated therein can cause any of the reaction indicated in regulation 13 of these Regulations then the waste shall be considered as a dangerous waste as described in paragraph (2) of this regulation.

(4) The following methods shall be used by the Agency to determine, representative samples as waste, that is –

(a)   crushed or powered material – ASTM standard method (D346-75);

(b)   extremely viscous liquid – ASTM standard method (D140-70);

(c)   flash like material – ASTM standard method (D22234-76);

(d)   solid or rock like materials – ASTM standard (D1452-65);

(e)   solid or rock like materials – ASTM standard method (D420-69);

(f)   containerized liquid waste (COLIWASA) described in "Test methods for the evaluation of solid waste, physical/chemical methods SW-846 USEPA (1985);

(g)   liquid waste in pits, ponds, lagoons and similar reservoirs – "Pond sampler" described in "Test Methods for the evaluation of solid waste, physical/chemical methods" SW – 846, USEPA (1985).

15. – (1) a solid waste is pits is ignitable if its representative sample   ignitability test has any of the following properties, that is -

(a)   it is liquid, other than an aqueous solution containing less than 24 per cent alcohol by volume and has a flash point of less than 60^0C as defined by Pensky-Mertens Close cup Tester using the test method specified in ASTM standard D-93-79 or D-93-3278;

(b)   it is not a liquid and is capable, under standard temperature and pressure, of causing fire through friction, absorption of moisture or spontaneous chemical changes and, when ignited, burns so vigorously and persistently that it creates a hazards;

(c)   it is an ignitable compressed gas;

(d)   it is an oxidizer.

(2)   An ignitable solid waste not designated as dangerous waste under any of the Agency's Dangerous Waste List or the Agency's Dangerous Waste Criteria shall be assigned the dangerous waste number FDOO1.

16. – (1) A solid waste is corrosive if its representative sample has   corrosivity
any or more of the following properties, that is -   testing

(a)   it is aqueous and has a (PH) less than or equal to 2 or greater than or equal to 12.5 as determined by a (pH) meter using method 5.2 in Test Method for the evaluation of solid waste, (Physical/Chemical methods);

(b)   it is a liquid and corrodes steel (5AE 1020) at a rate greater than 6.35mm or per year at a test temperature of 55^0C as determined by Standard Test Method 01-69 as standardized in "Test Methods for the evaluation of solid waste, (Physical/Chemical methods);

(c)   it is a solid or semi-solid, and when mixed with an equal weight of water results in a solution, the liquid portion of which has the property specified in such paragraph (a) of this regulation.

(2)   A corrosive solid waste not designate as a dangerous waste under any of the agency Exclusive List of Dangerous Substances listed in Schedule 12 to these Regulations or has any of the characteristics of Dangerous waste Criteria

listed in regulation 13 of these Regulations shall be designated DW and shall be assigned the Dangerous Waste number FD-002.

17. – (1) A solid waste exhibits the characteristic of reactive if a representative sample of the waste has any of the following properties, that is -

Reactive
Test

(a)   it is normally unstable and readily undergoes violent changes,; without detonating;

(b)   it reacts violently with water;

(c)   it forms potentially explosive mixtures with water;

(d)   when mixed with water, it generates toxic gasses, vapours or fumes in a quantity sufficient to pose danger to human or animal health or the environment

(e)   it is a cyanide or sulfide bearing waste which when exposed to (pH) conditions between 2 and 12.5 can generate toxic gasses, vapours of fumes in a quantity sufficient to present a danger to human or animal health or the environment.

Solid waste exhibiting characteristics waste of toxicity

18. Where a solid waste exhibiting the characteristic of EP toxicity is not designated a dangerous waste under any Agency's Dangerous Waste List or Dangerous Criteria, it shall be assigned to the Agency's dangerous waste number specified in Schedule 9 to these Regulations which corresponds to the toxic constituent causing it to be dangerous.

Generic Dangerous waste

19. A waste which exhibits any of the dangerous waste characteristics lists in Schedule 8 to these Regulations shall be assigned the dangerous waste number corresponding to the characteristic numbers exhibited by the waste as shown in Schedule 9 to these Regulations.

APPENDIX IV

FEDERAL ENVIRONMENTAL PROTECTION AGENCY ACT
(CAP. 131 LFN)

National Environmental Protection (Pollution Abatement in Industries and Facilities Generating Wastes) Regulations 1991

Commencement 5th August, 1991

In exercise of the power conferred upon me by section 37 of the Federal Environmental Protection Agency Act and of all of other powers enabling me in that behalf, I, Major Mamman Tsfo Kontagora (rtd), hereby make the following Regulations:

1. No industry or facility shall release hazardous or toxic substances into the air, water or land of Nigeria's ecosystems beyond limits approved by the Agency

Restriction on the release of toxic substances. Monitoring Pollution units

2, An industry or a facility shall –

    (a)   have a pollution monitoring unit within its premise;

    (b)   have on site a pollution control;

    (c)   assign the responsibility for pollution control to a person or body corporate accredited by the Agency.

3.   A discharge, including solid, gaseous and liquid waste from any industry or facility shall be analyzed and reported to the nearest office of the Agency every month, through a Discharge Monitoring Report.

4. An unusual discharge or accidental discharge of waste from any industry facility shall be reported to the nearest office of the Agency not later than 24 hours of the discharge.   *unusual or accidental Discharges*

5. An industry or facility shall submit to the nearest office of the Agency   *List of Chemicals*

   (a)   a list of the chemical used in the manufacture of its product;
   (b)   details of stored chemicals and storage conditions;
   (c)   where chemicals are bought, sold or obtained, the name of any secondary buyer.

6. The State and Zonal Offices of the Agency shall serve as Pollution Response Centres for co-ordinating pollution response activities   *Pollution Response Centre*

7. AN industry or a facility shall have a contingency plan approved by the Agency against accidental release of pollutions.   *Contingency Plan*

8. – (1) An industry or a facility shall set up a machinery for combating pollution hazard and maintain equipment in the event of an emergency   *Machinery for combating pollution etc*

(2)   An industry or a facility shall, for the purposes of paragraph (1) of this regulation, have a stock of pollution response equipment which shall be readily accessible and available to combat pollution hazards in the event of accidental discharge.

9. If there is a case of pollution emergency, the nearest office of the Agency shall serve as an "on-the-scene co-ordinator" to co-ordinate all response activities   *Pollution emergency*

Storage treatment and transport of harmful toxic waste

10. – (1) No person or body corporate shall engage in the storage of transportation of harmful toxic waste within Nigeria without a permit issued by the Agency

(2) The permit shall be in such form as may be determined by the Agency

Generator's Liability

11. – (1) The collection, treatment, transportation and final disposal of waste shall be the responsibility of the industry or facility generating the waste.

(2) An industry or a facility shall be liable for any clean-up, remediation or restoration connected with the waste and where necessary, compensation to all affected parties.

Industrial Layouts in each State

12. – (1) Each State of the Federation shall –

(a) designate industrial layouts which shall be separate from Residential areas: and

(b) provide buffer zones between industrial layouts and residential areas.

(2) A buffer zone shall =
(a) be rigidly kept away from developers;
(b) be monitored to prevent developing encroachment by developers

Strategies for waste reduction

13. An industry or facility including those to be established after the commencement of these Regulations, shall adopt in-plant waste reduction and pollution prevention strategies

Restriction on new source of pollution

14. – (1) NO new industry of facility shall commence production without compliance with the provisions of these

Regulations.

(2) The Agency shall prevent an industry or facility from commencing operation where the Agency believes that such industry or facility may constitute a new point source of pollution

Permissible Limits of Discharge into public drains, etc

15. – (1) No effluent with constituents beyond permissible limits shall be discharged into public drains, rivers, lakes, sea or underground injection without a permit issued by the Agency or any organization designated by the Agency

(2) No oil, in any form, shall be discharged into public drain, rivers, lakes, seas, or underground injection without a permit issued by the Agency or any organization designated by the Agency.

(3) Application for a permit and the permit shall be in Forms set out in the schedule to these Regulations or as specified by the Agency.

(4) The Agency may revoke a permit issued under paragraph (1) of this regulation if the Agency is satisfied, after the due enquiry, that the industry or facility has not complied with any of the conditions specified in the permit.

(5) Revocation of a permit shall be in the Forms F and D set out in the Schedule to this Decree or as specified by the Agency.

(6) An industry or a facility with a new point source of pollution or a new process line with a new point source shall apply to the Agency for discharge permit not later than 80 days before commencing the discharge of any effluent arising from any operation.

16. – (1) Solid wastes generated by any industry or facility, including sludge and all bye-products, resulting from the operation of pollution abatement equipment, shall be disposed of in an environmentally safe manner

Solid wastes to be disposed in environmental safe manner

(2) No industrial solid waste shall be disposed of in any municipal; landfill.

17. An industry or a facility which is likely to release gaseous, participate, liquid or solid untreated discharges shall install, into as system, appropriate, abatement equipment in such manner as may be determined by the Agency

Release of gaseous matters

18. The surroundings of a factory or facility shall be maintained to preserve their aesthetic and sanitary conditions.

Surrounding of factories

19. No industry shall expose an employee to any hazardous condition in his place of work.

Safety of workers

20. The Forms set out in the Schedule to these Regulations shall be used for the purposes specified therein with or without modification by the Agency

Environmental impact assessment, etc

21. The Agency shall demand environmental audit from existing industries and environmental impact assessment from new industries and major development projects and the industries shall comply within 90 days of the receipt of the demand.

22. A person or body whether corporate or unincorporated who contravenes any provision of these Regulations shall be guilty of an offence and liable on conviction to the penalty specified in section 35 or 36 of Act

Penalty

23. These Regulations may be cited as the National Environmental Protection (Pollution Abatement in Industries and Facilities Generating Waste) Regulations 1991

Citation

APPENDIX V

FEDERAL ENVIRONMENTAL PROTECTION AGENCY ACT (CAP. 131 LFN)

National Environmental Protection (Effluent Limitation) Regulations 1991

Commencement: 15th August, 1991

In exercise of the powers conferred upon me by Section 37 of the Federal Environmental Protection Agency Act and of all of other powers enabling me in that behalf, I, Major Mamman Tsofo Kontagora (rtd), hereby make the following Regulations:-

1. – (1) Every industry shall install anti-pollution equipment for the detoxification of effluent and chemical discharges emanating and chemical discharge emanating from the industry.

(2) An installation made pursuant to paragraph (1) of this regulation shall be based on the Best Available Technology (BAT), the Best Practical Technology (BPT) or the Uniform Effluent Standards (UFS

2. – (1) The selected waste water parameters for the industries specified in column 1 to these Regulations are set out in columns 2 and 3 respectively of the Schedule

(2) The parameters shall be continuously monitored to ensure compliance with these Regulations.

3. – (1) An industry which discharges effluent shall treat the effluent to a uniform level as specified in Schedule 2 to these Regulations to ensure assimilation by the receiving water into which the effluent is discharged

(2) The nearest office of the Federal Environmental Protection Agency shall be furnished from time to time with the composition of any effluent treated as specified in paragraph (1) of this regulation.

4. An industry specified in column of Schedule 3 to these Regulations shall be subjected to the additional sectoral effluent limitations set out in columns 2 and 3 respectively of the Schedule

Addition sectoral effluent limitation treatment.
Penalty

5. A person who contravenes a provision of these Regulations is guilty of an offence and liable on conviction to the penalty specified in section 35 or 36 of the Federal Environmental Protection Agency Act

6. These Regulations may be cited as National Environmental (Effluent Limitation) Regulations 1991

Citation

APPENDIX VI

EXCERPT FROM DECREE NO. 86

(10th December, 1992)   Commencement

THE FEDERAL MILITARY GOVERNMENT hereby decree as follows:

PART – GENERAL PRINCIPLES OF ENVIRONMENTAL IMPACT ASSESSMENT

1. The objectives of any environmental impact assessment (hereafter in this Decree referred to as "the Assessment") shall be.

(a)   to establish before a decision taken by any person, authority, corporate body or unincorporated body including the Government of the Federal, State or Local Government intending to undertake or authorize the undertaking of any activity that may likely or to a significant extent affect the environment or have environmental effects on those activities shall first be taken into account;

(b)   to promote the implementation of appropriate policy in all Federal Lands (however acquired) States and Local Government Area, consistent with all laws and decision making processes through which the goal and objectives in paragraph (a) of this section may be realized.

(c)   To encourage the development of procedure for information exchange, notification and consultation between organs and persons when proposed activities are likely to have significant environmental effects on boundary or trans-state or on the environment of bordering towns and villages.

REFERENCES

Adegoroye, Adegoke (1997), "FEPA and other Environmental Agencies: An Understanding of the laws of Environment, the Role of NGO and Expectations of FEPA from the Nigerian Institute of Public Relations"; Paper presented at the Annual General Meeting and Conference of the Nigerian Institute of Public Relations (NIPR), held at Hotel Presidential, Port-Harcourt, 26th November.

Aina, A. (1990) Harmful Waste (Special Provision, Decree 42, of 1988", *Nigerian Business Law and Practice*, Vol. 2, No. 2.

Akpa, M.N. (1995): "Newspaper Coverage of Urban Waste and Environmental Sanitation in Nigeria: A Content Analysis of Four National Daily Newspapers". Unpublished Thesis, Enugu State University of Science & Technology.

Aliede, J.E. (2000), "Environmental Public Relations and the Ecological Imperatives of the New Millennium, *Journal of Public Relations Management,* Vol. 1, No. 2, p.21

Anokwute, N.C. (1999) "Newspaper Coverage of Environmental Pollution in Nigeria: A Content Analysis of Four Nigerian Newspapers", M.Sc Research Project, Public Relations Programmes, Department of Marketing, University of Nigeria (Nsukka), Enugu Campus.

Anon, (1989) *Environment and Urbanization* Vol. 1, Number 1, Cable News Network (CNN) (2003) UDS, November 11.

Anon. (1999), "Learning How to Restore the wilds of Eden," *Time International magazine, (October 14).*

Anon. (1992), *Urban Age Journal,* Vol. 1, NO. 1. September

Anon. (1994), *Journal of Environmental Planning and Management,* Vol. 31, No. 3.

Babbie, Earl R. (1973), The Practice of Social Research, Belmont, California; Wadaworth Publishers.

Baker, Viccy (2002), "Public Relations and the Challenges of Democratic Governance in Africa: The Southern African Case study", paper presented at the conference and Annual General Meeting of the Nigerian Institute of Public Relations (NIPR), held on June 26-28th at the Gateway Hotel Abeokuta, Ogun State.

Bernay, Edwards L. (1923), *Crystallizing Public Opinion*

Bernay, Edwards L. (1986), *The Later Years: Public Relations*

Bertalanffy, Ludwig Von (1968), *General Systems Theory Foundations, Development and Applications* New York: George

Braziller, Inc. (Revised Edition)

Bitner, John (1997) *Introduction to Mass Communication* (5th ed.)

Black Sam (1990) *Introduction to Public Relations,* London: Modino Press.

Borstin, Daniel (1962), *The Image: A Guide to Pseudo Events in America,* London: Weindenfield and Nicholson.

Brookes, S.K. and A.C. Jordan, R.H. Kimber and J.J. Richardson (1076), "The Growth of the Environment as a Political Issue and Britain", *British Journal of Political Science,* No. 6.

Bruce, E. (1993): "Green Communications in the Age of Sustainable Development", *Gold Paper No. 9,* IPRA.

Budd, Richard (1961): "U.S. News in the Press Down and Under", *Public Opinion Quarterly,* No. 28.

Budd, Richard et al. (1967): *Content Analysis of Communications. New York: Macmillan Company.*

Canfield Bertrand and H. Frazier Moore (1973), *Public Relations Principle, Cases and Problems,* Homewood, Illinois: Richard D. Irwin.

Canfield, Bertrand and H. Moore (1977), *Public Relations: Principles, Cases and Problems,* Hoomewood, Illiois, Richard D. Irwin. Inc. p.50.

Carlson, Roberts O. (1973) "The Role of Research in Public Relations," in *Lesly's Public Relations Handbook* Vol. 1 (2nd Ed.), London: Prentice Hall, Inc.

Chafetz, Janet S. (1978). *A Primer on the Construction and Testing of Theories in Sociology.* Ithaca, Illinois: F.E. Peacock Publishers Inc.

Chokor, B.A. (1985). Reported in E. Ebinne (2002), "Diffusion of Ecological Information: An Analysis of the Performance of Selected Print Media", Research Project, Public Relations Programme, Dept. of Marketing, University of Nigeria (Nsukka), Enugu Campus, Enugu.

Commoner, B. (1972), *The Closing Circle: Compounding the Environmental Crisis:* London; Jonathan Cape Publishers.

Cutlip Scott M. Allen H. Center and Glen M. Broom (1984), *Effective Public Relations* (7th ed.) New Jersey: Prentice Hall Inc.

Cutlip Scott M., Allen H. Center and Glen Broom (1985), *Effective Public Relations,* (7th ed.) New Jersey: Prentice Hall.

Danielson, Wayne (1958): "Content Analysis in Communication Research" in Ralph Nafziger and David White (eds), *Introduction to Mass Communication Research,* Louisianna: Louisianna State University Press.

Dominick, J. and Wimmer R.D. (2000) *Mass Media Research: An Introduction,* Belnot: Wadsworth Publishing Coy.

Drucker, Peter (1973), *Top Management,* London: Hienneman, Publisher.

Ebinne, Emmanuel Ikpai (2002).

Edeani, David (1993), "Public Relations, Public Opinion and Attitude Change" in A.O. Salu (ed), *Public Relations for Local Government in Nigeria.* Lagos: talkback Publishers Limited, p.109.

Essaghah Arthur A. and Adibe Ernest C. (1999), *Environmental Impact Assessment in Nigeria: Principles, Procedures and Practice* Vol. 2, Enugu, Immaculate Publications Ltd.

Etuk, E. (1997) "El Nino to Cause Food Shortage", *Guardian,* December, 16.

Festinger, Leon (1957), *A Theory of Cognitive Dissonance, Evanston: Row and Peterson Publisher.*

Fisher, Auberry B. (1978), *Perspectives on Human Communication,* New York: Macmillan.

Gruning James E. and Todd Hunt (1984), *Managing Public.*

Harrison, Paul (1993) *The Third Revolution: Population, Environment and Sustainable World,* London: Penguin Books.

Harrison, Paul (1993), *The Third Revolution Population, Environment and Sustainable World,* London, Penguin Books Limited.

Hart, Jim (1961): "The Flow of International News into Ohio, *Journalism Quarterly,* Vol. 59.

Hennesey, Bernard C. (1981), *Public Opinion* (4th ed.) Monetary, California: Books and Cole Publishing Company.

Hester, Al et al (ed). (1982), *Handbook for Third World Journalists,* Georgia (USA): The Center for International Mass Communication Training and Research. Miller, Tyler G. (1995), *Environmental Sciences: Sustaining the Earth,* California: Wadsworth Publishing Company.

Holgate M.N. (1993), "A Perspective on Environmental Pollution. "in Hardman, Mc Eldoweney and White (eds) *"Pollution : Ecology and Biotreatment ",* England: Longman Group, U.K. Ltd, p.3.

Holsti, Ole R. (1996), *Content Analysis in the Social Sciences and Humanities,* Massachusets: Addison Wesley Publishing Company.

Houghton, R. (1994) *Global Warming: The Complete Briefing,* New York: Cambridge University Press.

Igben, H.G.O. (1998), "Public Relations and Control of Community Hospitalities in the Nigerian oil industry" unpublished Research Project, Department of Marketing, University of Nigeria, Enugu Campus.

Ikeagwu, Egbui K. (1998) *Groundwork of Research Methods and Procedures,* Enugu Institute (for) Development Studies, University of Nigeria.

Ikechukwu E. Nwosu transposed (1996), *Public Relations Management: Principles, Issues and Applications,* Lagos: Dominican Publishers.

Insights **(1959-1986),** New York: H & M Publishers.

IPRA (1992), *Code of Practice on Public Relations and the Environment, IPRA* Publications, London.

Jacob, M. (1991), *The Green Economy: Environment, Sustainable Development and the Policies of the Future,* London Pluto Press.

Jefkins, Frank (1985), *Marketing Management, Advertising and Public Relations,* London: Macmillan Publishers.

Jones, R.R. and Tom Wigley (eds) (1989), *Ozone Depletion: Health and Environmental Consequences,* New York: John Wiley and Sons Limited.

Kamena, K.W. (1991), "Plastics, Packaging and the Environment: A U.S.A Perspective", Paper Presented to the Conference Organized by Papra Technology and Piro International on "Is Plastics Packaging Rubbish?", London, 30th Dec. to 1st January.

Kelly, Joseph (1993): "Reporting the Environment", *Journal of Land and Water,* Vol. 2, No. 15.

Kerlinger, F. (1973), *Foundations of Behavioural Research,* New York: Holt, Rinehart and Winston, Inc.

Kotler, Philip (1977), *Marketing Management,* New York: Prentice Hall Inc.

Kotler, Philip (1993), *Marketing Management:* New Delhi: Prentice Hall of India.

Kotler, Philip (1994), *Marketing Management: Analysis, Planning, Implementations and Control* (7th ed.) New York: Prentice Hall Inc.

Lerner, Daniel (1958), *The Passing of Traditional Society,* Glencoe Illinois. The Free Press.

Lesley, Philip (1978), *Lesley's Public Relations handbook,* (2nd ed.) London: Prentice Hall inc., P.56.

Lipman, Walter (1922): *Public Opinion* McMillan Press.

Mabbutt, J.A. (1991) "A New Global Assessment of the Status and Trends of Desertification", *Environmental Conservation,* Vol. 11, No. 2.

Maslow, Abraham (1954) *Motivation and Personality,* New York: Harper and Row, Inc.

Mba, Chike H, (2002), "Environmental Management for Rural Development: Strategies, Techniques and Implementation", Paper presented at the National Workshop on New Strategies in Environmental Management Protection, Rural Centre, Port-Harcourt, 24th – 26th July.

Mba, H.C. and J. Ogbazi (1979) "Urban Planning Perspective and Emerging Concepts", in Mba et al (1979) *The Principles and Practice of Urban and Regional Planning in Nigeria,* Awka,; Mekslink Publishers.

Mbithi, J. (1971). *African Religions and Philosophy,* New York: Praeger Publishers.

McCarthy, E.J. (1982), *Essentials of Marketing,* Homewood, Illinois: Richard D. Irwin, Inc.

McCombs, M. and D. Shaw (1973): "The Agenda Setting Hypothesis", *Journalism Quarterly,* Vo. 3

McGuire, William J. (1989). "Theoretical Foundations of Campaigns". In Roland E. Rice and Charles Aikins (Eds) *Public Communication Campaigns.* Newbury Park: Sage Pblications.

Miller G.T. (Sr) (1998), *Living in the Environment* (10th ed.): Cincinnati, Ohio: Wadsworth Publishing Co.

Miller, G.T. (Jr) (1989), *Living in the Environment,* Tenlted, Cincinati: Wadsworth Publishing Company, p. 16.

Mitchell, A. and L. Levy (1989) "Green About Greens", *Marketing,* 14th September.

Mody, Bella (1991). *Designing Messages for Development Communication: An Audience Participation Based Approach.* London: Sage Publications.

Moemeka, Andrew (1991). "Perspectives on Develoment Communication". In S.T. Kwame Boafo (Ed.) *Module on Development Communication No. 1.* Nairobi: African Council for Communication Education (ACCE).

Moffat, Ekoriko (1995), *Newswatch,* December, 18

Muchall, D. (1992) "Environmental management: The Relationship Between Pressures Group and Industry: A Radical Redesign", in D. Koechlin and K. Muller (eds.) *Green Business Opportunities,* London: Pitman Limited.

Nelson, Ridley (1988), *Dryland Management: The Desertification Problem,* Washington, D.C: World Bank.

Nkpong, S.J. (1994), "Global and Nigerian Environmental Problems Analysis", SIRF, Calabar 3rd November

Nkwocha; Jossy (1999) *Media Relations Management,* Lagos: Zoom Lens Publishers.

Nwankwo N. and Ifedi C.N. (1980) "Case Studies on the Environmental Impact of oil production in Nigeria" in Sada, P.O. and F.O. Odermerho (eds), *Environmental Issues and management: Proceedings of the National seminar on Environmental Issues and Management in Nigerian Development,* Ibadan: Evans Press.

Nwosu, Ikechukwu E. (1990) *Mass Communication and National Development,* Aba, Frontier Publishers.

Nwosu, Ikechukwu E. (1986), "Mobilizing Peoples' Support for Development in Africa", *Africa Media Review,* Vol. 1, No. 1.

Nwosu, Ikechukwu E. (1990) "Vital Communication and Public Relations Principles and Techniques for the Modern Manager/Administrator," in Ikechukwu E. Nwosu (ed) *Mass Communication and National Development,* Aba; Frontier Publishers.

Nwosu, Ikechukwu E. (1991). Towards Effective Application of Public Relations, Public Opinion and Attitude Change Theories and Techniques in Nigeria's Local Government Administration and Management, *Nigerian Journal of Marketing,* Vol. 4, No. 13.

Nwosu, Ikechukwu E. (1992) "Disseminating Information on Environmental Issues and Problems in Rural Africa". Towards an Integrated Model", *Mass Media and the Environmental in Africa,* Nairobi: African Council for Communication Education (ACCE).

Nwosu, Ikechukwu E. (1992), "Public Relations: An Introduction to the Principles, Functions and Practice" in Ikechukwu E. Nwosu and S.O. Idemili (eds) *Public relations: Speech, Media Writing and Copy,* Enugu: ACENA Publishers.

Nwosu, Ikechukwu E. (1993), "Disseminating Information on Environmental Issues and Problems in Rural Africa", Towards an Integrated Model," in S.T. Kwame Boafo (ed.), *Media and Environment in Africa: Challenges for the Future,* Nairobi: ACCE Publications.

Nwosu, Ikechukwu E. (1994) The Newspaper in the Development of Developing Nations", in Andrew Moemeka (ed) *Communicating for Development: A New Pan-Disciplinary Perspective,* New York: University of New York Press.

Nwosu, Ikechukwu E. (1995), *Mass Media and African Wars: Media Images of Crises in Africa,* Enugu: Star Publishing and Company Limited.

Nwosu, Ikechukwu E. (1996) *Public Relations Management: Principles, Issues and Applications,* Lagos: Dominican Publishers.

Nwosu, Ikechukwu E. (1996), *Effective Media writing,* Enugu: Precision Publishes.

Nwosu, Ikechukwu E. (1997) "Media Images of Environmental Issues in Nigeria: Implementations for Public Relations Managers", Paper presented at the NIPR Conference (AGM), Presidential Hotel, Port-Harcourt, 26th November, (Unpublished in Image Maker, Vol. 1, No. 2, 1995).

Nwosu, Ikechukwu E. (2000) *Marketing Communication Management and Media: An Integrated Approach,* Lagos: Dominican Publishers.

Nwosu, Ikechukwu E. (2000), "Case Study on Environmental Incidents and Public Relations", Paper Presented at the Intensive Workshop on Public Relation in the Oil Industry Organized by the SPDC and BEEC at Gans-Court Hotels Benin, June, 2000.

Nwosu, Ikechukwu E. (2001) *Marketing Communications Management and Media: An Integrated Approach,* Lagos: Dominican Publishers.

Nwosu, Ikechukwu E. (2002) "Mass Media and Grassroot Develoment in Nigeria: An Overview", Invited Paper presented at the workshop on Community Reporting held at Zodiac Hotel, Enugu, August, 29.

Nwosu, Ikechukwu E. (2002) "Service Quality and Customer Expectations", in U.JIF. Ewurum (ed.) *Managing Service Quality in the Nigerian Public Sector,* Enugu: Smart-Link Publishers.

Nwosu, Ikechukwu E. (2002), "Environmental Public Relations Management: Implementation Models, Strategies and Techniques" Invited paper presented at the National Workshop on Strategies for Environmental Management and Protection, held at the Integrated Cultural Centre, Port-Harcourt, July 24-26.

Nwosu, Ikechukwu E. (2003) "Managing Small and Medium Scale Enterprises in a Competitive and Depressed Economy" in J.O. Omah and I.E. Nwosu (eds) *Empowering Small and Medium Scale Enterprises in Nigeria,* Enugu: ECCIMA.

Nwosu, Ikechukwu E. (2003), "Public Relations, the Environment and Sustainable Development; Expanding Our Knowledge base and Skills", The 2003 Sam

Epelle Memorial Lecture, Organized by the Nigeria Institute of Public Relations (NIPR), Makurdi, Nigeria, November, 2004.

Nwosu, Ikechukwu E. (2003), "Public Relations, the Environment and Sustainable Development in Africa: Expanding and Applying our Knowledge Base and Strategic Options", Invited paper presented at the 16th Sam Epelle Annual Memorial Gold paper lectures held at the Aminu Isa Kontagora Arts Theatre Complex Makurdi, Benue State (Nigeria), December, 11.

Nwosu, Ikechukwu E. (2003), Environmental Public Relations Management: "Implementation Models, Strategies and Techniques", *The Nigerian Journal of Communication*, Vol. 2, Nos 1&2.

Nwosu, Ikechukwu E. (2004) "Managing Small and Medium Scale Enterprises in a Competitive and Depressed Economy", in Julius Onah, Ikechukwu Nwosu & O. Ocheoha, (eds) *Empowering Small and Medium Scale Enterprises in Nigeria,* Enugu: CIDJAP Press.

Nwosu, Ikechukwu E. (2004) "Managing Small and Medium Scale Enterprises in a Competitive and Depressed Economy", in J.O. Onah, I.E. Nwosu and O. Ocheoha (eds) *Empowering Small and Medium Scale Enterprises in Nigeria,* Enugu: ECCIMA & MANMARK.

Nwosu, Ikechukwu E. (2004), "Environmental Public Relations Management and Sustainable Development", *Nigerian Journal of Marketing.* Vol. 5, No. 1.

Nwosu, Ikechukwu E. and U. Ekwo (eds) (1996), *Mass Media and Marketing Communication: Principles, Perspective and Practices,* Enugu: Thought Communication Publishers, p.11

Nwosu, Ikechukwu (1995): *Mass Media and African Wars: Media Images of Crises in Africa,* Enugu: Star Publishers.

Nwuneli, Onuora and Opubor, Alfred E. (Eds) (1988), *Communication and Human Needs in Africa.* New York: Blind Beggar Press.

Nwuneli Onuorah (1985) *Mass Communication in Nigeria:* A Book of Readings, Fourth Dimension Limited.

Obeng-Quaidoo, Isaac (1986). "A Proposal for New Communication Research Methodologies in Africa," *Africa media Review,* Vol. 1, No. 1.

Odo, Ozongwu (1999) *Research in Social and Behaviroural Sciences,* Enugu: XYZ Publishers P.42.

Okigbo, Charles (1992). "Integrated Marketing Communication: The New Advertising", Paper presented at the Nigeria Union of Journalists (NUJ) Conference on Communications in the 21st Century, Enugu, September 9-11.

Okorie, E. (1992) Cited in Uchegbu, Smart (1998), *Environmental Management and Protection,* Enugu: Precision Printers and Publishers.

Olakunori, O.K. (1997), *Successful Research ESUT:* Enugu:

Olugbemi, K.S.O. (1992), Newspaper Coverage of Environmental Issues and Problems in Nigeria: A Case Study of Lagos State", Unpublished Thesis, University of Lagos.

Olukoya, Sam (1995) *Newswatch,* December, 18.

Omole, Sola (1997) "The Environmental and Operations: Oil Industry Perspective" Paper presented at the NIPR Conference/AGM, Hotel, Presidential Port-Harcourt, 26th November.

Omole, Sola (1998), "The Environment and Operations; Oil Industry Perspectives", *IMAGE MAKER NIPR Journal,* Ogun State, Vol. 1, No. 1.

Onumonu, Ane (1986) "Environmental Issues and Corporate *Public Relations",* *Public relations in Africa,* Vol. 1, No. 1.

Onyeador, Stanley O. and Ikwuegbu Ngozi M. (1999), *Environmental Impact Assessment,* Enugu, Frank Miller Publishers.

Osuala, E.C. (1991), *Introduction to Research Methodology,* Onitsha: African-Fep Publishers, p.18.

Ozongwu, Maurice O. (1992) *Guide to Proposal Writing in Social Sciences and Behavioural Sciences,* Enugu: SNAAP Press.

Peattie, Ken (1995) *Environmental Marketing: Meeting the Green Challenge:* London: Pitman Publishing Ltd.

Porrit, Jonathan (1991) *Save the Earth,* Great Britain: Dorling Kindersley Limited.

Rattan, Lal (1985) "Soil Erosion and Productivity in Tropical Soils", in S.A. Ebuwafey (ed) *Soil Erosion and Conservation,* Iowa: Conservation Society of America.

Raymond, Simon (1980), *Public Relations: Concepts and Practices,* Columbia, Ohio: Grid Publishing Company.

Scissors, Jack (Ed). *Integrated Marketing Communications,* Evanston: University of Illinois Press.

Simon, E. (1991) "Marketing Green Products in the Triad", *Columbia Journal of World Business,* Vol. 27, Nos. 3 & 4, pp. 268-285.

Smith, L. Graham, (1993), *Impact Assessment and Sustainable Resource Management,* New York, Longman Group UK Ltd.

Smith, P.R. (2000), *Marketing Communications: An Integrated Approach* (2ⁿᵈ ed.), London: Kogan Page Limited.

Smut, J. (1920), *Holism and Evolution,* London: UN World Commission on the Environment.

Sobowale, Idowu (1986): "Content Analysis: A New Perspective", Paper presented at the ACCE Workshop on Communication research, held in Harare, Zimbabwe, October 13-18 1986.

Taylor, A.E. (1963) "Critias, 111, b-d" in Plato: *Collected Dialogues,* Princeton University Press.

Tenant, T. and M. Campanale (1991) "A Long Term Investment", *Environmental Strategy* 1991, London: Campden Publishing.

The New Partnership for Africa's Development (NEPAD), A Publication of JDPC & CIDA, 2002.

The Oxford English Dictionary, **Vol. Vii, (1978**) London: Oxford University Press, p.1081.

This Day **(2002**), July 2.

Uchegbu, Smart N. (1998) *Environmental Management and Protection,* Enugu: Precision Printers and Publishes.

Uffoh, Vincent (2000) "An Evaluative Study of the Application of the RICEE Model in Environmental Sanitation Management: A Case Study of Enugu State", Unpublished M.Sc. Research Project; Department of Marketing, UNN, Enugu Campus.

Uffoh, Vincent O. (2002) "An Evaluation of the Public Relations RICEE Model as a Strategy Towards Combating the Menace of Environmental Degradation in Enugu Metropolis," Unpublished Masters Degree Project, Public Relations Programme, University of Nigeria, Enugu Campus.

Ugochukwu, Onyema (2002) "Environmental Audit: A Strategy for Managing Environmental Problems in the Niger Delta, Paper presented at the 3-Day National Training Workshop on "New Strategies in Environmental Management Protection", Integrated Cultural Centre, Port Harcourt, July 24 – 26.

Ukpong, S.J. (1994) "Global and Nigerian Environmental Problems Analysis" SIRF, Calabar, November, 3.

Umeh, L.C. and Smart N. Uchegbu (1997), *Principles and Procedures of Environmental Impact Assessment (EIA),* Lagos: Computer Edge Publishers.

Umeh, Louis G. and Smart N. Uchegbu (1997). *Principles and Procedures on Environmental Impact Assessment* (EIA) Lagos: Amazing Grace Printing and Publishing Company.

UNEP (1988), *Environmental Impact Assessment: basic Procedures for Developing Countries* New York: UNEP Publications.

United Nations Environmental Programme (1990), *The State of the Environment:* Children and the Environment, New York, UNICEF.

Unterman, J. (1994), "America Finance: Three views of Strategy *Journal of General Management,* Vol. 3, (1987), *Our Common Future,* New York: WECD Publications.

Vesilind, P.A. and J.S. Pierce (1977) *Environmental Engineering.* England: Ann Arbor: Science Publications.

Walker, Clive (1992) "To Adventure is to live," *Safari,* Vol. 9. No. 2.

Wiebe, Gerhart (1953) "Some Implications of Separating Opinions from Attitudes," *PUBLIC OPINION QUARTERLY* Vol. 17.

World Bank (1978), *Environmental Considerations for the Industrial Development Sector,* Washington, D.C.: World Bank Publications.

INDEX

www.ingramcontent.com/pod-product-compliance
Lightning Source LLC
Chambersburg PA
CBHW060327200326
41519CB00011BA/1860